David James Hamilton

On the pathology of bronchitis, catarrhal pneumonia, tubercle and allied lesions of the human lung

With illustrations

David James Hamilton

On the pathology of bronchitis, catarrhal pneumonia, tubercle and allied lesions of the human lung
With illustrations

ISBN/EAN: 9783337140670

Printed in Europe, USA, Canada, Australia, Japan

Cover: Foto ©berggeist007 / pixelio.de

More available books at **www.hansebooks.com**

ON THE

PATHOLOGY OF BRONCHITIS, CATARRHAL PNEUMONIA, TUBERCLE,

AND ALLIED LESIONS OF THE HUMAN LUNG.

BY

D. J. HAMILTON, M.B. F.R.C.S.E. F.R.S.E.

PROFESSOR OF PATHOLOGICAL ANATOMY (SIR ERASMUS WILSON CHAIR),
UNIVERSITY OF ABERDEEN

With Illustrations

London:
MACMILLAN AND CO.
1883

The Right of Translation and Reproduction is Reserved

TO

SIR ERASMUS WILSON,

AS A SLIGHT TRIBUTE TO HIS MAGNANIMITY IN PROMOTING

PATHOLOGICAL SCIENCE,

THIS LITTLE WORK IS RESPECTFULLY

𝔇𝔢𝔡𝔦𝔠𝔞𝔱𝔢𝔡

BY THE AUTHOR.

PREFACE.

THE little work which I now offer to the profession had its origin in a series of papers which appeared in *The Practitioner* for the years 1879 and 1880 Since then I have carefully revised all the observations and made such alterations as bring the subject-matter up to the present day. I am well aware that many points contained within the book will excite adverse criticism on the part even of those who have made Tubercle and Phthisis subjects of special study. In working out the materials for it, however, I have gone on the principle of forming an unbiased opinion according to the facts before me, and if I have thereby rendered clearer a subject which at the present day must present a somewhat chaotic aspect to the Practitioner of Medicine or Student of Pathology I shall consider that my object has been in great part accomplished.

To Dr. T. L. Brunton, in his capacity of Editor of *The Practitioner*, I have to return my sincere thanks for his general superintendence of the articles.

D. J. HAMILTON.

THE UNIVERSITY,
  ABERDEEN,
    January 27, 1883.

CONTENTS.

PART I.—ON BRONCHITIS.

	PAGE
Introductory	1
Structure of normal bronchi	3
Mucosa	3
Epithelium	4
Basement membrane	10
Inner fibrous coat	11
Muscularis	12
Outer fibrous coat	12
Mucous glands	15
Regeneration of epithelium in normal bronchi	17
Origin of the mucous corpuscles	20
Acute congestion of the bronchi	21
Acute bronchitis	24
Chronic bronchitis—varieties	44
Chronic bronchitis following an acute attack	44
Chronic bronchitis following valvular lesion of heart	60
Chronic bronchitis due to inhalation of foreign matters	64
Chronic bronchitis associated with chronic disease of the kidney	68
On chronic interstitial pneumonia as a complication of bronchitis	69
On vesicular emphysema and collapse of the lung as complications of bronchitis	89
On bronchiectasy as a complication of bronchitis	95
On catarrhal pneumonia as a complication of bronchitis	99

PART II.—ON CATARRHAL PNEUMONIA AND TUBERCLE IN THE HUMAN LUNG.

	PAGE
Introductory	101
The structure of the wall of the natural air-vesicles	102
Effects of increased blood-pressure suddenly applied to the blood-vessels of the lung	110
On the effects of long-continued excessive blood-pressure in the lung	129
Catarrhal pneumonia. First stage	136
Catarrhal pneumonia. Second stage	147
Tubercle in the human lung	162
Primary tubercle of lung	164
Secondary tubercle of lung	193
Secondary tubercle of lung accompanying cirrhosis	194
Mode of development of secondary tubercle accompanying cirrhosis of the lung	203
The degenerations of tubercle	205
Catarrhal pneumonia. Third stage	211
On a peculiar form of catarrhal pneumonia which is liable to be mistaken for tubercle	219
On the supposed contagiosity of tuberculosis and pulmonary phthisis	224

LIST OF ILLUSTRATIONS.

FIG.		PAGE
1.	Section of normal human lung	4
2.	Section of normal human bronchus	5
3.	Cells from deepest and middle layers of bronchial epithelium	7
4.	Surface view of bronchial epithelium stained with nitrate of silver	7
5—7.	Cells from deepest layer of bronchial epithelium	8
8—11.	Different transitional stages in the formation of columnar epithelium	9
12.	Surface view of bronchial mucous membrane	10
13.	Basement membrane and attached epithelial cells of a bronchus	10
14.	Peri-bronchial fibrous tissue	13
15.	Peri-arterial fibrous tissue	16
16.	Peri-arterial lymphadenoid deposit from cat	16
17.	Partially and completely developed columnar cells	18
18.	Transverse section of part of mucous membrane ; acute bronchitis	28
19.	Transverse section of small bronchus ; commencing acute bronchitis	29
20.	Transverse section of entire bronchus ; acute bronchitis	31
21.	Peri-bronchial fibrous tissue ; acute bronchitis	38
22.	Section of a mucous gland ; acute bronchitis	39
23.	Section of a mucous gland ; acute bronchitis	40
24.	Section of a bronchial ganglion ; acute bronchitis	42
25.	Transverse section of a bronchus in a state of chronic catarrh	47
26.	Mucosa of bronchus seen in a state of chronic catarrh	48
27.	Bronchial cartilage undergoing atrophy	57
28.	Small abscess at mouth of a mucous gland	59
29.	Transverse section of a bronchus in regurgitant mitral disease	62
30.	Bronchus—coal-miner's lung	65
31.	Particles of inhaled dust in coal-miner's lung	66
32.	Chronic interstitial pneumonia	73

LIST OF ILLUSTRATIONS.

FIG.		PAGE
33.	Air-vesicles in same	75
34.	Partially obliterated artery in same	77
35.	Wall of a bronchiectatic cavity in same	81
36.	Chronic interstitial pneumonia	87
37.	Scheme of same	88
38.	Small bronchus in acute bronchitis, occluded by a plug of inhaled catarrhal secretion	93
39.	Alveolar cavities from lung of kitten, stained with nitrate of silver	103
40.	Same—profile view	104
41.	Same; showing blood-vessels injected with nitrate of silver	103
42.	Section of a coal-miner's lung, showing the lymphatic spaces of the alveolar wall injected with pigment particles	109
43.	Section of an alveolus from the lung of a person who died from mitral disease	132
44.	Acute catarrhal pneumonia	138
45.	Catarrhal cells—acute catarrhal pneumonia	139
46.	Acute catarrhal pneumonia	141
47.	Acute catarrhal pneumonia. Blood-vessels injected	143
48.	Acute catarrhal pneumonia	145
49.	Catarrhal pneumonia, second stage	150
50.	Catarrhal pneumonia, second stage	151
51.	Primary tubercle of lung	170
52.	Fully developed tubercle of lung	173
53.	Giant-cell from centre of a tubercle of the lung	174
54.	Large oval giant-cell from tubercle of lung	174
55.	Giant-cell from tubercle of lung, with inhaled particles of carbon in its interior	175
56.	Remains of a giant-cell in process of fibrous transformation	176
57.	Primary tubercle of lung. Developing	181
58.	Enlarged connective-tissue corpuscle from a sarcoma	190
59.	Giant-cell, from a myeloid tumour	191
60.	Secondary tubercle of lung	201
61.	Secondary tubercle of lung	204
62.	Catarrhal pneumonia, third stage	216
63.	Catarrhal pneumonia, third stage, showing obliteration of an artery	217
64.	Disseminated catarrhal pneumonia	221
65.	Primary tubercle of lung	222

PART I.

ON BRONCHITIS.

INTRODUCTORY.

THE catarrhal affections of the bronchi are so familiar to every one, from their common occurrence, that a knowledge of what takes place in the structures affected, how their mechanism is injured, and why it is that recovery is often so tedious and very frequently incomplete, becomes a matter of the greatest interest and importance to the practising physician. There is perhaps no class of diseases that one is more called upon to treat than that of the catarrhal affections of mucous membranes generally, and, curiously enough, there is perhaps no class of lesions whose pathology we know less about. There are almost no detailed researches as to what takes place in such mucous membranes, and we have, consequently, no means of knowing accurately whether we have to do with an exalted normal secretion, an abnormal secretion, an affection of the mucous glands, or a lesion of the mucous surface itself. Where does the catarrhal secretion come from? What is it made up of? And why does the catarrhal inflammation in some cases tend to pass off and allow the mucous membrane to resume its healthy function, while in others, and these too commonly, it assumes a chronic form in which there is the greatest difficulty in bringing about a favourable result? These are the practical questions which

I shall keep before me in the present inquiry, and the object of what follows will be to show what structures are more particularly involved, how it is that recovery takes place, and why in chronic cases recovery is so protracted and frequently incomplete.

As in the study of any morbid process, we find that the primary departure from the normal to that which we understand as a diseased state is merely a matter of degree, either an exaggeration of, or a deficiency in, the normal standard, it is imperative before investigating what we call morbid processes, to become familiar with the natural histological structure and histogenesis of the part concerned. Disease is merely a departure from what we find to be the average normal condition, and before we can interpret what the disease signifies we must be thoroughly conversant with the means by which the organ or tissue is formed, grows, and has its waste elements repaired. The key to the elucidation of a diseased state is to be found in the observation of what occurs in the discharge of the natural functions of the part.

It will, therefore, be necessary before entering upon the subject of this paper to consider somewhat carefully the structure of the parts that we shall have to deal with, and, along with this, the manner in which their different textural constituents are reproduced; and as it is particularly with the *mucous membrane* of the bronchi that we shall have to do, our observations will, in the first place, be directed to its structure with more especial regard to its epithelial covering. Strange as it may appear, we know comparatively little of how laminated epithelial tissues generally are reproduced. They are constantly being shed, and although we may conjecture that the deeper layers give rise to the more superficial, yet the manner in which this is accomplished, and the source of the deepest layer of all, are still not clearly demonstrated. It is evident that, unless we understand this, we can never appreciate any morbid overgrowth or degeneration to which they are liable, and we shall see that the study of the whole subject of the catarrhal affections of mucous membranes is founded on this previously acquired knowledge. Why it should happen, again, that the bronchi and alveoli in the same lung are lined with totally different epithelium, is also matter for consideration. Both are formed from the same germinal layer in the embryo, so that the fact of the one being columnar and the other squamous, cannot rest on their primary source of origin. It is clearly an afterwork, suitable for the different purposes which the epithelial covering has to subserve in the economy of the part which it protects. Keeping in mind, however, that they were originally both derived from the hypoblast, we may better perceive how in reality there is not so very wide a difference between them as might at first sight appear, the one being perhaps a further development of the other.

Structure of the Normal Bronchi and their Surroundings in Man.

After the trachea divides into its two primary branches, the latter very soon subdivide dichotomously until the ultimate bronchioles are reached, which, finally, open into the infundibula and air-vesicles. If one of these larger bronchi be laid open in a healthy person who has been suddenly killed, the mucous membrane which lines it is seen to have a pinkish colour, and is covered by a coating of frothy mucus. When this mucus is detached, the underlying epithelial surface is seen to be glossy and of a fine velvet-like consistence. It is comparatively transparent, so that in certain places, where it is thin, the underlying inner fibrous coat may be seen shining through it. If such a bronchus be cut transversely, it is found to be made up of at least three coats, probably a fourth, the innermost being the *mucosa* itself, the intermediate one being the *muscularis*, and the outer one the *adventitious* or *outer fibrous coat* containing the elliptical cartilages.

The Mucosa.—In the middle-sized bronchi it is thrown into longitudinal folds, but in the smaller and smallest it forms a

STRUCTURE OF THE BRONCHI.

smooth membrane covered by epithelium (Fig. 1, a). It is composed of three layers, comprising the inner fibrous coat lying next to the muscularis, a basement membrane covering and attached to this, and an epithelial covering superficial to all, with numerous glands for the secretion of mucus. We shall examine the epithelial covering first.

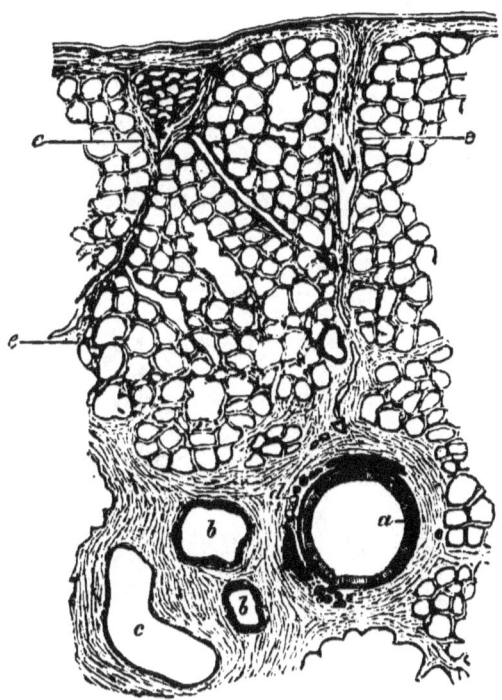

FIG. 1.—Section of normal human lung— × 50 Diams. (reduced ¼). a, small bronchus; b b, branches of pulmonary artery; c, branch of pulmonary vein; d, air-vesicles; e, interlobular septa.

On careful examination with a magnifying power of about 480 diameters it can be easily recognised that the character of the cells in the epithelial layer is by no means the same at different depths, but that while the superficial cells have a columnar shape, those underlying them are transitional in character (Fig. 2); while deeper than this still there

STRUCTURE OF THE BRONCHI.

is a perfectly flat layer lying on a homogeneous basement membrane, more like the endothelial cells seen on serous surfaces. A general view of the structure of a small bronchus is given in Fig. 2, where the stratified character of the epithelium will also be noticed. The broad end of each columnar

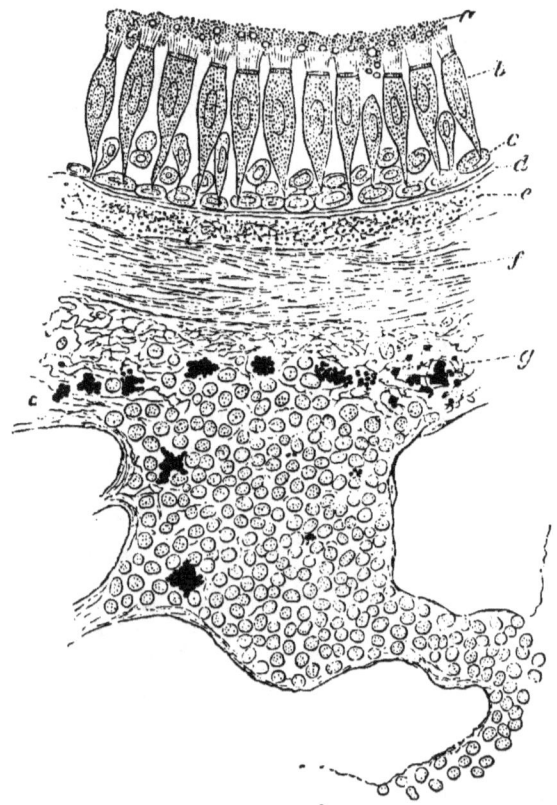

Fig. 2.—Section of normal human bronchus— × 450 Diams. *a*, precipitated mucus on surface of epithelium; *b*, columnar epithelial cells; *c*, deep germinal layer of cells (Debove's membrane); *d*, elastic basement membrane; *e*, inner fibrous coat; *f*, muscularis; *g*, outer fibrous coat with lymphadenoid deposit in it, and pigment granules.

cell has a fringe of cilia attached to it, and these usually bear on their surface a little viscid mucus (Fig. 2, *a*). Each of them has a nucleus, but the surrounding body of the cell is frequently so granular that this may be hidden from view. The underlying attachments of the columnar cells can be most easily studied in bronchi which have been the subject of acute congestion for from ten to twenty hours. The effect of the congestion is to cause a partial desquamation of the epithelium, so that the cells become isolated from one another, and, if such a bronchus be hardened in dilute solution of chromic acid, and the mucous surface either simply scraped or fine sections made from it and examined in water, the underlying attachments of the columnar cells, which otherwise are usually torn across, may be readily studied. It can then be noticed that the columnar cell does not usually end in a free tapering extremity, but has attached to it one of the rounded or somewhat flattened cells of the deepest stratum. This is best seen where the columnar cell is young and where its underlying attachment is thick and not so easily broken as when it becomes older and more attenuated (Fig. 3, a). Letzerich has figured this appearance (*Virchow's Archiv*, vol. liii. p. 493), but believes that the underlying flattened cell is a connective tissue corpuscle. That this is incorrect will be seen from the sequel, but it is apparent that what is represented in his figure is the above described attachment of a columnar cell to one of a different character underlying it.

Beneath the stratum of columnar cells there are two layers of a different character. The lower of these when looked at from above seems to be flat and to correspond very much to what Debove (*Comptes Rendus*, 1872; *Archives de Physiologie*, 1874) has described as an endothelium underlying the epithelium proper in the trachea, bladder, and intestine. In Fig. 4 is represented a surface view of a portion of the lower part of the trachea of a rat from which the columnar cells had been partially removed by pencilling with a camel's-hair brush, and then stained with nitrate of silver. At the upper part of the figure is seen the free surface of the columnar cells as they appear when silvered with a magnifying power of 450 diameters. They are comparatively small, and the openings of chalice cells can be noticed here and there between them. At the lower part of the figure, however, the superficial columnar epithelium has been brushed off, bringing this layer into view. It will be seen that its cells are large and usually six-sided, differing slightly in size, some of them being smaller and evidently younger than the others. Although they appear to be flat when looked at from above, yet they are seen, on perpendicular section of the mucous membrane, to have a nucleus in the

Fig. 3.—Cells from deepest and middle layers of bronchial epithelium— × 450 Diams.

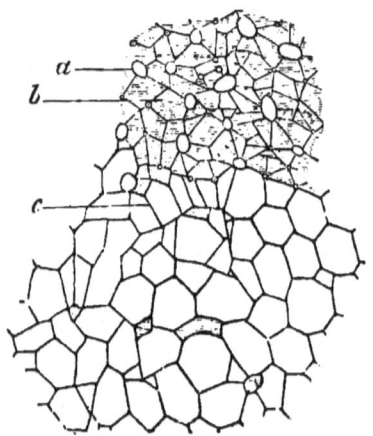

Fig. 4.—Surface view of bronchial epithelium stained with silver— × 450 Diams. *a*, free ends of columnar cells with openings of chalice cells; *b*, divisions between individual cells; *c*, deeper flat layer of epithelium (Debove's membrane).

centre which gives them an oval shape. In Fig. 5 is represented this layer in a chromic acid preparation scraped off from the mucosa of a bronchus in man. It is clearly made up of a pavement-like stratum of extremely delicate cells, some of them, in a state of germination, having apparently fallen out. On perpendicular section also of the mucous membrane, especially where there is commencing catarrh, this flat layer can be seen lying immediately in contact with the basement membrane to be afterwards described (Fig. 2, *d*).

It has been suggested by Debove (*loc. cit.*) that it is the analogue of the somewhat similar flat endothelial-like covering of the pulmonary alveoli, and certainly both its appearance in the normal bronchus and its behaviour under inflammatory stimulation would lend very considerable support to this theory.

Between this flat layer of epithelium and the columnar cells on the surface there is an intermediate stratum composed of cells in form more or less transitional between the deepest flat layer and the most superficial columnar. They are pyriform or *battledore* shaped, and have their tapering extremities implanted among the tesselated cells just noticed,

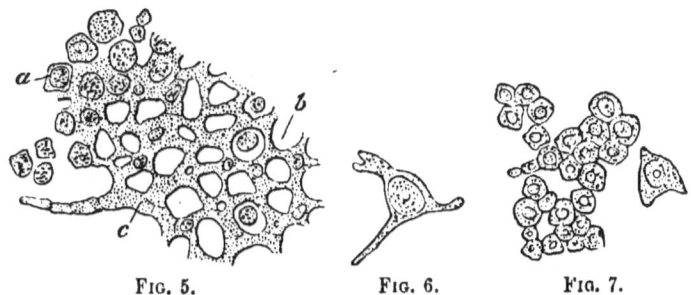

Fig. 5.   Fig. 6.   Fig. 7.

Cells from deepest layer of bronchial epithelium— × 450 Diams. *a*, germinating epithelial cell ; *b*, cavity formed by one of these falling out ; *c*, endothelial-like markings.

while the blunted end of the cell extends upwards between the ordinary columnar epithelium (Fig. 2). They have the same granular appearance and large nucleus and vacuole of the cells in which they are implanted, and they differ materially in size, some of them looking like small buds (Fig. 3, *b*), others being longer and forming bodies like those delineated in Figs. 9 and 10, *a*. The blunted round extremity which they usually possess is not invariably present, for it is sometimes noticed that the broad part of the cell has become constricted in the middle (Fig. 10, *b*), and has developed a nucleus in each segment. Further, in some rare cases where the cell is young, it presents the appearance seen in Fig. 11, where the part of the cell nearest the end of attachment is broad, while that in front of it is prolonged into a sharp point, and, frequently, as in the figure, several processes come off from it and serve to form an attachment to the more rounded underlying cells.

It is, therefore, clear that in the epithelial lining of a bronchus we have not only a layer of columnar cells, but that there are distinctly seen at least two strata of a different character underlying this, the intermediate stratum being composed of cells of transitional shape, while deepest of all there is a layer of flat cells almost like an endothelium. The pointed extremities of the columnar cells,

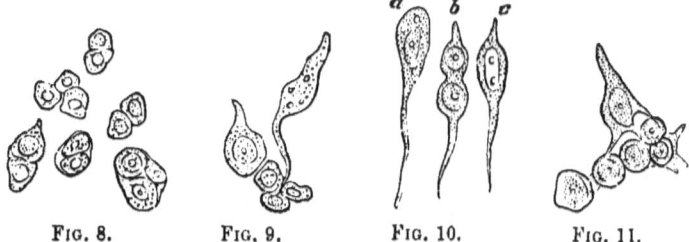

Fig. 8.   Fig. 9.   Fig. 10.   Fig. 11.
Different transitional stages in the formation of columnar epithelium— × 480 Diams.

especially in such as are of recent formation, run downwards to be implanted in the deepest stratum, and are connected with actively dividing cells contained within it; and it will afterwards be seen that a knowledge of this relationship and the fact of there being a deep stratum of actively germinating cells below those of a columnar shape, are most important preliminaries in the elucidation of the changes produced in acute and chronic bronchitis.

When the smallest bronchi are reached, immediately before they enter into the infundibula and air-vesicles, these stratified layers of epithelium are lost, and a single somewhat cubical covering is substituted, and, finally, on entering the air-vesicles the cells become very much flattened and extremely delicate in contour, resembling in every respect the endothelium of a serous surface. It would therefore appear as if the structural type of the epithelium covering the whole of the air-passages, from the trachea downwards, was flat, and that the occurrence of the columnar cells in the bronchi is merely an accessory to the underlying squamous layer.

The next point to which I have directed especial attention in this investigation is the question as to what the epithelial covering of the bronchial mucous membrane rests upon. Does it lie free on the inner elastic coat, or is there something interposed between them? In Fig. 12 is represented a detached portion of the mucous membrane of a bronchus in man, looked at from above, which has been slightly squeezed

Fig. 12.—Surface view of bronchial mucous membrane— × 450 Diams. *a*, columnar ciliated cell; *b*, germinal epithelial cell from deep layers; *c*, elastic fibre from inner fibrous coat.

Fig. 13.—Basement membrane and attached epithelial cells of a bronchus— × 450 Diams. *a*, germinating epithelium; *b*, elastic basement membrane.

out by the pressure of a cover-glass. The columnar cells (*a*) are seen flattened and having their tapering extremities implanted in the germinal layer (*b*); while deeper than this still are seen the elastic fibres of the inner fibrous coat shining through. Looked at in this manner it would appear as if there were nothing intervening between the deeper layers of epithelial cells and the underlying elastic fibres. When, however, a section is made perpendicular to the surface of the bronchial mucous membrane, there is always seen *in man*, more especially in the large and middle-sized bronchi, a perfectly homogeneous boundary layer (Fig. 2, *d*), having a

STRUCTURE OF THE BRONCHI.

wavy outline staining of a pinkish colour with carmine, possessed, apparently, of no fibrous or cellular structure, and not acted on by strong acetic acid.

Silver staining, so far as I have seen, fails to bring out any further structure in this membrane; and other re-agents, such as perosmic acid, likewise fail to develop anything further in it than a perfectly homogeneous translucent substance. In order, if possible, to detach this homogeneous basement membrane from its underlying attachments, I spread out a bronchus as flatly as possible on a piece of cork, and fixing it in this position, cut it layer by layer from the surface downwards in the freezing microtome. The first section removed was made up of the epithelium, but at the second section a layer was detached having the appearance represented in Fig. 13. A group of germinating epithelial cells is seen at one part (*a*), clearly showing that I had got down to the deeper part of the epithelial covering, while underlying this is a perfectly homogeneous membrane thrown into wavy folds (*b*). A few elastic fibres of the inner fibrous coat were seen underlying this, but they are omitted to prevent complicating the drawing. The membrane appeared quite hyaline, and no endothelial markings or nuclei were visible upon it, a few granules, here and there, being the only interruption to its invariable homogeneous translucency. It reminded one very much of the structureless appearance of the matrix of hyaline cartilage, and, in other respects, it seemed to be a membrane acting a merely mechanical part in giving attachment to the epithelial covering, and separating it from the underlying vessels and elastic fibres of the inner fibrous coat. This comes to be a most important structure in imparting the superficial character to the catarrhal affections of the bronchi; and, further, we shall find that as it becomes œdematous in such lesions, its structure and connections can be very much better made out. I call this layer of mucosa the basement membrane.

The inner fibrous coat (Fig. 2, *e*) follows next in order, and it is this which gives to the deeper part of the mucous membrane its fibrous character. It is mainly composed of elastic fibres with a few bundles of colourless wavy fibrous tissue, and it is abundantly supplied with branches from the bronchial artery, which form a ramifying plexus immediately under and projecting into the basement membrane just described. It also contains many branched connective tissue spaces, similar to those seen in the peri-arterial fibrous tissue of the pulmonary artery. The surface of the bundles of fibrous tissue is covered by an endothelium, whose nuclei are seen as large oval and extremely granular-looking bodies. The lymphatic system of the inner fibrous coat can be demonstrated by brushing off the epithelium and then silvering. It consists of a branching network of lymphatic vessels and capillaries inclosing strands of a rounded, oval or irregular shape, in which is seen a close plexus of irregularly-shaped plasma spaces. These latter undoubtedly form the radicles of the peri-bronchial lymphatic system, and they correspond, no doubt, to those seen in the walls of the pulmonary air-vesicles.

The muscularis (Fig. 2, *f*) is that coat which lies next to the inner fibrous layer, and, in some respects, may be said to belong more properly to the mucosa than to be reckoned as a separate investment. It is composed of widely interlacing bundles of non-striated muscular fibres between which numbers of lymphatic vessels ramify.

The outer fibrous coat or adventitia is of very great importance in the investigation of the changes produced in acute and chronic bronchitis. Its structure is that of somewhat loosely disposed bundles of white fibrous tissue with intermingled elastic fibres, arranged concentrically to the lumen of the bronchus, and passing continuously into the adventitious coat which surrounds the wall of the larger branches of the pulmonary artery (Fig. 1). On tracing the connections of this adventitious fibrous layer outwards into the lung substance, a further most important relationship can be noticed (Fig. 1). Every here and there, passing in from the deep layer of the pleura, coarse fibrous bands can be seen

STRUCTURE OF THE BRONCHI.

which separate the one pulmonary lobule from the other, and are consequently known as the lobular or interlobular septa. When these are followed inwards, it is apparent (Fig. 1) that after giving off several branching processes, they ultimately become connected to the adventitious coat of the bronchi and large branches of the pulmonary artery; and, as they run in from all sides, more especially at the periphery of the lung, there can be seen on a transverse section of a bronchus, perhaps three, four, or more such processes coming off at different parts of the bronchial wall. In none of the

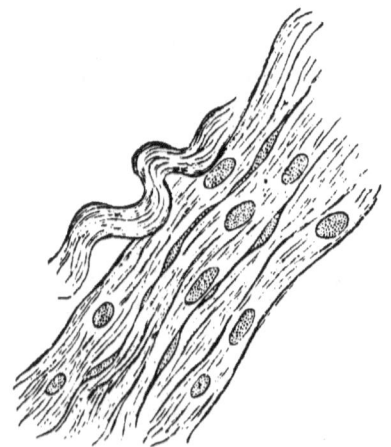

Fig. 14.—Peri-bronchial fibrous tissue— × 450 Diams.

mammalia which I have examined are the adventitious coat and lobular septa so well developed as in man, and we shall see that this relationship is a most important one in the formation of bronchiectatic cavities in old-standing cases of bronchitis.

The bundles of fibres forming the adventitious coat of the bronchus and artery are covered by flat connective tissue corpuscles, whose nuclei (Fig. 14) when seen on their surface, are of an oval shape, but on section, have a spindle shape, from the bulging of their centre. Branched spaces of considerable size intervene between these bundles of fibres (Fig. 15), spaces in which, in the human lung, carbon particles are invariably met with, filling them partially or completely. They seem to be the lymphatic spaces of the peri-bronchial tissue. On tracing them by means of the carbon particles which they contain (and which form, as in the coal-miner's lung, a natural injection of the lymphatic system) further outwards into the substance of the lung, they are found to have a free communication with those similar spaces which lie in the fibrous tissue surrounding the pulmonary artery, and apparently pour their contents into them; while these, again, are continuous with lymphatic vessels of large size seen in the lobular septa. The lymphatics of the lobular septa lead out to the deep layer of the pleura, where an abundant plexus of large lymphatics ramifies, and ultimately, no doubt, they pass to the bronchial glands at the base of the lung. The superficial layer of the pleura *never*, so far as I have seen, becomes pigmented, the lymphatics in this, representing as they do those of the pleura proper, having, apparently, few or no communications with those of the deeper layer.

I shall notice these relations of the lymphatics in the human lung subsequently, but this introduction will be sufficient to aid in the description of the changes they undergo in acute bronchitis. The lymphatic glands, in the human lung, are continued along the bronchi to a considerable distance within the lung substance, and they are afterwards represented in the bronchioles by the peri-bronchial lymphadenoid deposits situated in the adventitia (Fig. 2). In man, these peri-bronchial lymphadenoid deposits have an indistinct reticular structure, and, apparently, lie on the side of a lymph path, as in the corresponding structures in the lower animals; for, almost invariably, inhaled carbon-particles can be seen passing in a row by the side of such a deposit, while some of them have become entangled within it (Fig. 2). While these peri-bronchial lymphatic deposits are occasionally met with, the corresponding peri-arterial formations, so well developed in the lung of most of the lower animals, are either absent in the healthy human lung, or are of quite a rudimentary character. In the cat, for instance, these peri-arterial deposits are made up of a distinct flat-meshed reticulum, with lymphoid cells contained in it (Fig. 16), very much the same as in a lymphatic gland, and have a distinct boundary line, apparently due to the distended peri-arterial fibrous tissue. In man, however, no such distinctly developed lymphatic-gland-like bodies are to be seen around the arteries, and the bronchi only possess them here and there. In Fig. 2, g, one of these lymphadenoid deposits is seen in connection with a human bronchus. It is placed in the adventitious coat, and extends for some distance along the walls of the neighbouring pulmonary alveoli. There is no distinct reticulum visible, but this may be due to the mass of cellular structures comprised within it. I have never been able, however, to see the same flat-meshed reticular structure in these human peri-bronchial formations that can be observed in the corresponding peri-arterial structures in the lower animals; and it is evident that although the former are undoubtedly of the same nature, they are much less highly organized than those found in many other mammalian lungs.

The mucous glands are found in those bronchi which are provided with cartilages. They are racemose in character, and they run partly in the interspaces between the cartilages, and partly lie on the inner surface of these structures, where they have a flattened appearance. They each have a duct which opens into the bronchus by a trumpet-shaped orifice (Fig. 1), while the neck of the gland becomes rapidly constricted. The epithelium lining the neck of the gland is a double layer, the superficial cells being columnar and ciliated, while those in the deep stratum are spheroidal or pavement-like. Near the termination of the duct of the gland the columnar cells become smaller in size, transitional in character, and finally disappear, the deeper germinal layer being alone continued into the alveoli of the gland. In this latter

STRUCTURE OF THE BRONCHI.

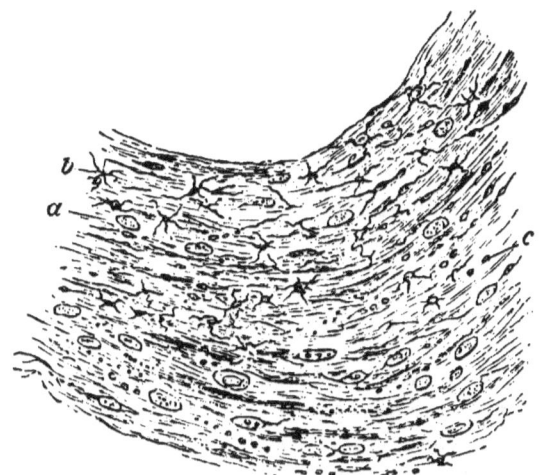

Fig. 15.—Peri-arterial fibrous tissue— × 450 Diams. *a*, connective tissue corpuscle; *b*, stellate lymphatic space.

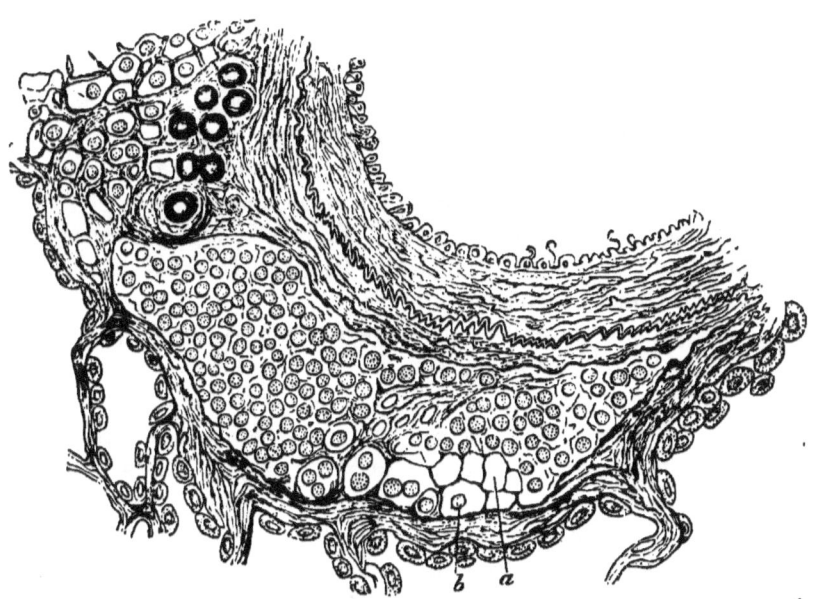

Fig. 16.—Peri-arterial lymphadenoid deposit from cat— × 450 Diams. *a*, lymphatic mesh-work; *b*, contained lymph corpuscle.

situation the majority of the cells are irregularly spheroidal or tesselated in character. They seem to be the continuation of Debove's layer in the trachea and bronchi. The whole course of the duct is provided with a finely fibrous basement membrane, and between the ramifications of the gland there are numerous blood-vessels, with a well-marked stroma, and numerous connective-tissue plasma spaces.

The Regeneration of the Epithelial Covering in the Normal Bronchi.

Having considered so far the minute structure of the normal bronchi, the next point for examination before I enter upon the consideration of the catarrhal affections of the mucous membrane, is how the epithelium, which is constantly being shed during health, is repaired; where the columnar cells come from, and how they are formed.

We have seen that in bronchi of large and medium size there may be said to be three varieties of epithelium arranged in so many superimposed strata, the whole of them resting on an apparently elastic structureless basement membrane. The layer which is next to the basement membrane, and consequently deepest in the series, is flat, and has more the character of an endothelium; that which succeeds it is of a transitional nature, while the most superficial is the layer of completely developed columnar cells. These various layers differ so much from each other, that one does not, at first sight, perceive how they can have any mutual relationship. In bronchi from the human subject, however, which have been slightly hardened in a weak solution of chromic acid, it can be seen that the one is only a further advanced development of the other, that the deeper layers are merely, as it were, the "raw material" from which the perfect columnar cell arises. The study is best made in bronchi which have been the subject of acute congestion a few hours before death, the resulting œdema of the basement membrane serving to produce partial desquamation or

18 REGENERATION OF THE EPITHELIAL COVERING.

loosening of the epithelium, and consequently to isolate the one cell from the other. The preparations are best made by scraping the mucous membrane gently and examining the detached cells simply in water, and then afterwards comparing the appearances procured in this manner with those exhibited on a perpendicular connected section of the mucosa.

In persons who have been killed by opium, certain of the bronchi are specially favourable for this study. With very little care it can be noticed that the columnar cells do not usually end in free pointed extremities as represented in Fig. 17, but that, as mentioned before, they are attached to an underlying flat and often polygonal cell. There can be further seen, lying between the columnar cells, *battledore* or tadpole-shaped cells having a similar deep attachment. Schulze has figured them in his article on the lung in Stricker's *Histology*, but he does not appear to have recognised their significance in relation to the underlying layer of germinating cells. If a portion of the intermediate transitional layer (Figs. 3 and 9) be detached, these pyriform offshoots can be seen sprouting from the flat-cell layer underneath, while the cells in this layer itself are all seen to be in the most active state of germination.[1]

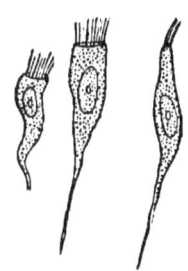

FIG. 17.—Partially and completely developed columnar cells.

Putting all these facts together, and after carefully studying

---

[1] The bodies represented by Klein, in his classical work on the *Lymphatic System of the Lung* in Fig. ix. A and B, as pseudostomatous cells, I believe to be merely transitional forms of columnar epithelium. They are much better seen in the human bronchus than in that of the guinea-pig from which his figure is taken, although essentially the same structures are to be found in the bronchial epithelium of all the mammalia that I have examined. The interepithelial bodies represented at B (*loc. cit.*) I would look upon as young epithelia, with the processes which are so often found in connection with them. The above-mentioned figure (ix. B) is surely somewhat diagrammatic, probably caused by the surroundings having been omitted.

the epithelial surface, both detached and *in situ*, I have arrived at the following conclusions as to the formation and repair of the columnar cells.

The cells of Debove's layer are those from which they are produced, and the first appreciable step in their formation is the division, of one of the polygonal cells contained in it, into two, four, or more, of smaller size (Figs. 7 and 8). The nucleus of the parent cell divides first, and this is followed by the fissiparous division of the body. Very soon, however, either while attached to the parent mass, or when removed to some distance from it, one of these elongates and becomes pyriform in shape, its broad end looking into the lumen of the bronchus, the other having a deeper attachment to the cells from which it originally sprung. It grows, gains in bulk, and develops a large oval nucleus with vesicular nucleolus (Fig. 3, *b*), and at times it is seen to have more than one delicate process of attachment to the underlying germinal mass (Fig. 11), the free extremity of the cell in such a case usually being the more pointed of the two, the other being somewhat blunted. All of these various processes are apparently lost at a later period, and then the cell assumes a pyriform shape, and it seems very probable that these cells, while the processes are still present, correspond with the pseudostomatous cells described by Klein in his work on the *Lymphatic System of the Lung*. This pyriform offshoot then stretches upwards between the columnar cells, and assumes a *battle-dore* or tadpole shape, and I think would then correspond to the bodies already referred to (p. 18) as figured by Schulze (*loc. cit.*). The next transformation is a very interesting one ; the blunt end of the *battle-dore-like* cell elongates into a sharp point (Fig. 10, *c*), and then this proboscis-like point can be seen to become longitudinally striated (Fig. 17). The end of the cell then dilates, and the consequent splitting up of the striated extremity gives rise to the cilia of the fully developed columnar cell (Fig. 17). The whole process of formation of the columnar cells seems quite rational when looked at in this way, for when we consider the common origin of the whole epithelium of the air-passages, we can see how the flat cell may be the original type, and how the columnar cell might be merely a further advance in evolution. Even although the epithelium of the air-vesicles seems so different from that of the larger bronchi, it only requires to be slightly stimulated, as in acute catarrhal pneumonia, for it to undergo division and germination. And if we look at such a surface a short time after the inflammatory excitation has commenced, pyriform or *battle-dore-like* processes will be seen to be given off from the flat proliferating cells of the air-vesicles identical with those sprouting from the deepest layers of the bronchial epithelial covering. With a very slight further advance it can easily be conceived how the development of a perfect columnar cell might be arrived at; and were this irritation continued for a lengthened period it can be understood how a certain impression might be made on the endothelial surface, resulting, in the course of time, in a permanent developmental peculiarity.

Origin of the Mucous Corpuscles.

One thing remains to be noticed at this stage of our inquiry, namely, the origin of the mucous corpuscles. In normal mucus they are seen to be round bodies, differing somewhat in size, but usually about that of a leucocyte; and, when a section is made perpendicularly through the mucous membrane, they are seen lying on the surface of the cilia imbedded in stringy mucus. On careful examination of the epithelial surface round or stretched-out cells of a similar nature can occasionally be seen, making their way between the columnar cells, and finally emerging on the free surface. They arise from the deeper layers, and are frequently formed in the manner shown in Fig. 10, *b*, where one of the pyriform cells from the deeper layer is seen becoming constricted in the middle, and a nucleus forming in each portion. By a further contraction complete separation of the two occurs, the peripheral segment making its way outwards, as a mucous corpuscle. I would look upon the mucous corpuscles, then, simply as the waste products of the germinal or formative layer of epithelium, which have not gone on to complete development, but which have been thrown off in an unfinished condition. There seems every reason to believe, and indeed it can be shown beyond doubt, that the same thing occurs in the mucous glands themselves, but that they are the exclusive source of the mucous corpuscles is certainly very far from the truth. The function of the mucous glands seems to be in great part, if not entirely, that of the secretion of the fluid element of the mucus, and in bronchitis this function is exalted to an extreme degree.

Acute Congestion of Bronchi.

Having described somewhat carefully the process by which the epithelial covering of the mucous membrane of the bronchi is reproduced in the normal condition, we are now prepared to consider the lesions which arise in connection with it. It will be more conducive to the understanding of the catarrhal processes if we first examine what occurs in the bronchus in acute congestion of a mechanical nature. For, although it is somewhat difficult to show where mere mechanical congestion ends and inflammatory congestion begins, or, in fact, to define what the difference between the two is in the early stages, yet I shall consider as mechanical those congestions which do not produce a continuous further change in the bronchial wall, but which, if the cause be removed, tend to resolution; and as inflammatory those which are accompanied by further changes of a secondary nature, resulting in the formation of what we know as inflammatory products. One of the best means of studying acute mechanical congestion in the human bronchi is presented in persons who die from opium-poisoning. The mucous membrane of the bronchi in persons who have met their death by this means is in an acutely congested state, and, as the

ACUTE CONGESTION OF BRONCHI.

subject of it has rarely lived beyond a few hours, we can calculate how much change has been produced in a given time. The bronchi also usually contain a considerable quantity of tough glairy mucus, a plug of which will very often be found in the glottis, obstructing it, and probably aiding in bringing about the fatal result. I have examined the bronchi of an otherwise typically healthy subject, aged twenty-five, who died in from ten to sixteen hours, so far as could be ascertained, after an overdose of opium, and found the following appearances :—In the smaller bronchi, which contained either very small cartilages, or none at all, there seemed to be no deviation from the normal state; the mucous membrane and surrounding lung tissue were quite intact, and the layer of columnar epithelium was perfectly preserved. There was, further, no marked congestion of these at the periphery of the lung, and all that was noticed on their free surface was the layer of mucus which is usually seen when they are quite normal. In the larger bronchi, however, an entirely different appearance was noticed, for here the whole bronchial wall, and more especially its inner fibrous coat, was in a state of acute congestion. The small arteries, veins, and capillaries were widely distended, and were all filled with blood corpuscles, and loops of the same could be seen running up to the basement membrane and projecting on its free surface. The latter structure was particularly evident, apparently on account of its having become very œdematous. It was clear and transparent, and was thrown into irregular folds on the surface, evidently from the infiltration of serous fluid into it. The columnar epithelium, here and there, was left unaltered; but throughout the greater extent of the surface of the mucosa it had been either partially or completely shed, so that the deeper layers alone were left adherent to the basement membrane. In this manner an extremely demonstrative view could be had of the structures in the deeper layers, for not only was the basement membrane œdematous and particularly easily seen on that account, but the different layers of cells were beautifully isolated on account of their partial desquamation. It might be argued, in opposition to what I have stated, and especially in regard to the cause of the desquamation, that these were, in part, *post-mortem* changes. In order to see what influence *post-mortem* changes might have on the epithelial surface of the bronchial mucosa, I instituted a series of experiments on the subject, with entirely negative results as regards the desquamation of the epithelium, an advanced state of *post-mortem* decomposition having apparently little effect in loosening the attachment of the columnar cells. In one experiment I placed the larger bronchi and part of the surrounding lung substance first in liquor amnii, in which it lay soaking for several days. I then removed it from this, and allowed it to be exposed for about a fortnight, at a warm temperature, until it was in an advanced state of putrefaction. On preparing and cutting this in the same way as the others, notwithstanding the long time which had elapsed since death, and the very trying *post-mortem* conditions under which it was placed, the epithelium was *perfectly preserved in situ*, and not the slightest desquamation of the superficial cells could be perceived. All my other experiments went to confirm this observation, and we shall see that the appearances above described as occurring in this acute congestion of the bronchi in opium-poisoning are also quite characteristic of the commencement of a bronchitis.

At the same time, however, that these changes were taking place in the vessels and epithelium of the mucosa, a further lesion was observed in its fibrous coat. This consisted in the accumulation of leucocytes in the interspaces of the fibres, an appearance that I shall have to notice more particularly in examining the lesions in acute and chronic bronchitis. I would therefore conclude that in a typically healthy subject, mere mechanical congestion, lasting only a few hours, is sufficient to give rise to great œdema of the basement membrane, and desquamation of the superficial layer of epithelium covering the larger bronchi. And, further, that coexistent with this there is noticed the accumulation of small round cells, apparently leucocytes, in the fibrous coat of the mucosa, which escape by diapedesis through the distended walls of the blood-vessels.

ACUTE BRONCHITIS.

From the fact of the mucous membranes of the trachea and larger and smaller bronchi being continuous, and having very much the same structure throughout, it might be expected that a catarrhal affection arising in the one would have a tendency to influence the other. And although, no doubt, this is true to a certain extent, yet it by no means necessarily follows that they are invariably all implicated simultaneously, and to the same extent. There are cases, for instance, where we have to do with a limited tracheitis, without the bronchi showing any particular signs of a similar affection; and what is still more usual, the smaller bronchi and air-vesicles may be the seat of extensive catarrh without the largest bronchi and trachea being co-existently implicated. The seat of the disease seems, for some unexplained reason, to be frequently limited to a certain portion of the respiratory mucous membrane, and if it does involve the other portions, does so only in a later stage of the disease, and after the vigour of the inflammatory process has expended itself on the primary seat. We accordingly find that it is almost possible to make a somewhat artificial division, according to the portion of the mucous membrane affected, into tracheo-bronchitis, where the lower part of the trachea and the larger bronchi are the main source of the catarrhal fluid, bronchitis proper, where the medium-sized bronchi are chiefly implicated, and capillary bronchitis where the smallest-sized bronchi are the chosen seat, the last being usually accompanied by more or less acute catarrhal pneumonia, due to the extension of the inflammatory process to the alveolar wall. The first is naturally less dangerous than the other two, and is usually met with in adults: the last, when occurring in children, is an extremely dangerous affection, on account of the narrow calibre and delicate structure of the parts where it has its source, and from the fact that it is always more or less complicated with an acute catarrhal pneumonia.

It is seldom that adults die from an attack of acute bronchitis alone; it is usually where a catarrhal pneumonia has been super-added that it proves fatal and that we have an opportunity of seeing what condition the bronchi are in. If the respiratory passages be examined in such a case, it will be found that the trachea contains a more than usual amount of catarrhal fluid, which has a yellow colour, and is viscid and frequently frothy. In the smaller bronchi the same secretion is noticed, but in greater abundance; and if they be squeezed, a little pellet of this yellow muco-purulent secretion is expressed. The colour of the fluid in the bronchi in such an acute attack, say of ten to fourteen days' duration, is invariably yellow, and of thicker consistence than the normal mucus. It is evident, therefore, that this is not a mere exalted normal secretion, but that it has altered from its usual greyish colour and slightly tenacious consistence, to a yellow, extremely viscid, more purulent-looking material. If the air-vesicles be at the same time involved, that is to say, if there is a concurrent pneumonia, it will be noticed, on looking superficially over the cut suface of the lung, that there are little patches of lung substance every here and there, which are slightly raised, and which are of a somewhat greyish-pink colour. They feel quite soft, and if the finger be rubbed lightly over the surface, they give much the same impression as that of a mass of frog's spawn. When these little masses are squeezed, a quantity of the same muco-purulent fluid exudes either from the cut surface, or from a small bronchus which may happen to communicate with the particular group of air-vesicles implicated. It is clear, therefore, that the whole character of the inflammation, whichever part of the respiratory tract may be affected, is what we understand as catarrhal, in which, instead of there being a solid substance, rich in fibrin, exuded on the surface, as in croupous pneumonia, or croupous bronchitis, there is a fluid secretion, containing instead a large percentage of mucin; and, further, the mere fact of the constant difference in the character of the morbid product in these two affections would lead us *à priori* to the supposition that they also differed in their manner of formation.

I have examined a great many cases of acute bronchitis arising either idiopathically or as sequelæ to other diseases, more notably measles and whooping-cough, but have found that in the whole of them the appearances are substantially the same; and hence the description of the one will, with a few modifications, correspond to that of the others. I have taken as my model, however, in the following remarks, an acute case, occurring in an otherwise healthy youth of nineteen years of age, who died from acute bronchitis, followed by acute catarrhal pneumonia; but it will be understood that the substance of what follows is drawn not merely from this case, but that all the points have been verified in several others. When we cut into such a lung, the bronchi and bronchioles are seen to contain much yellow muco-purulent fluid, which covers the surface of the mucous membrane, and can be pressed out in the form of a tough tenacious pellet. When this muco-purulent secretion is scraped off, what seems most remarkable is the immense congestion of the underlying mucous membrane, whose vessels can be seen ramifying on the surface. When the bronchus is cut across, the outer fibrous coat does not exhibit the same red colour, but contrasts visibly in its somewhat greyish-white glistening appearance with the congested mucosa. The smaller bronchioles may be completely choked up with the muco-purulent discharge, and the whole lung substance intensely congested, especially towards the larger bronchi. When the mucous membrane of the larger bronchi and the cut surface of the lung have been exposed to the atmosphere, the blood which they contain turns to a bright florid red or scarlet hue, differing markedly from what occurs under similar circumstances in mere mechanical hypostasis, where it remains of a more or less purple or brownish-red tint.

ACUTE BRONCHITIS.

In children who die from acute bronchitis, the naked-eye appearances are very much the same, with the exception that the bronchial secretion has usually accumulated in larger abundance in the bronchi of small size. All that we learn from the naked-eye examination is, that the mucous membrane is intensely congested, and is covered with this muco-purulent secretion. We cannot from such an examination tell what the causes of these appearances are, and therefore require to resort to microscopic examination to make these apparent.

It is always a matter of difficulty in man to get at the first change which ensues in the bronchi in acute catarrh. Most of the cases have advanced beyond this, and it is only by careful selection of certain bronchi that we are enabled to make this evident. On careful comparison, however, of many cases, I feel assured that the first deviation visible is a *relaxation and distension of the abundant plexus of bloodvessels ramifying in the inner fibrous coat*, immediately beneath the basement membrane, that is to say, of the branches of the *bronchial artery*. They become engorged with blood, so that on transverse section they appear like little cavities distended with blood-corpuscles (Fig. 19, *b*). In a few hours afterwards the basement membrane becomes much more apparent than it usually is, and, at the same time, more clear and homogeneous, while the surface is thrown into many folds (Fig. 19, *d*). These changes in the basement membrane are apparently due to its becoming œdematous, serous fluid being infiltrated into it from the underlying plexus of distended vessels; and we shall see that, as the acute irritation continues, this œdematous state of the basement membrane becomes a more and more well-marked feature. The next change, so far as I have been able to calculate, occurs in from twenty to thirty hours after the primary distension of the vessels, and consists in the loosening and desquamation of the columnar epithelium at the foci of greatest congestion (Fig. 19), some of the transitional forms of epithelium seen in the deeper layers being removed at the

28   ACUTE BRONCHITIS.

same time. This desquamation never seems to be general, there being usually in a transverse section of a bronchus one or two spots where the epithelium is still adherent. In some instances it takes place to an extreme degree, the whole epithelial covering being apparently detached. This, however, is rare, the columnar cells being those which are usually thrown off, leaving the more or less transitional layers

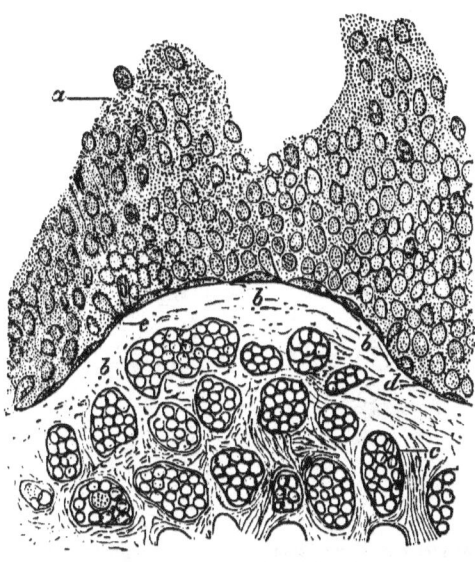

Fig. 18.—Transverse section of part of mucous membrane. Acute bronchitis : × 450 Diams. *a*, catarrhal secretion ; *b*, swollen basement membrane ; *c*, congested blood-vessels ; *d*, flat epithelial cell layer seen on section ; *e*, flat epithelial cell layer germinating.

adherent. The columnar epithelium is thus shed at a very early stage in the attack, and takes *no part whatever* in the afterchanges which ensue. It is never again seen until the other signs of acute inflammation, such as the distension of the vessels and œdema of the basement membrane, have passed off. Subsequently we shall see that it is gradually reproduced. It does not proliferate, there is no endogenous division of its nuclei, and it seems to behave very much in

ACUTE BRONCHITIS.   29

the same way as the *formed* layer of epidermis does in an inflammatory affection of the skin. It merely desquamates, and the only further change noticed in it is of a retrogressive nature, in that it undergoes fatty degeneration. It is, no doubt, partly destroyed by this means and partly expectorated; or some of it is inhaled into the smaller bronchi, where it can frequently be seen lying in large detached masses among the other catarrhal products. When it undergoes fatty degeneration, the protoplasm of the cells becomes granular and is resolved into oil globules and protein particles, which are

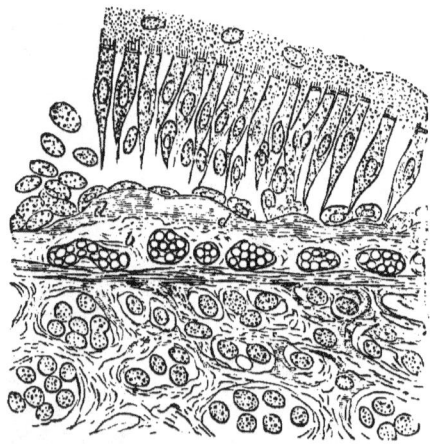

FIG. 19.—Transverse section of small bronchus. Commencing acute bronchitis: × 480 Diams. *a*, flat epithelial cell layer seen on section; *b*, inner fibrous coat; *c*, outer fibrous coat; *d*, basement membrane becoming œdematous.

expectorated or absorbed at a later stage of the disease. The fact of its not being reproduced during the course of the catarrhal inflammation corresponds with the clinical observation that *the expectoration in the later stages of acute bronchitis does not contain many columnar cells.* The cause of this desquamation of the columnar epithelium seems to be the œdema of the basement membrane loosening its underlying attachments, very much in the same way as the vesicles which form in an acute inflammatory affection of the skin loosen the attachments of the superficial layers of epidermis. The removal of this protective covering from the mucous membrane naturally leaves the latter in an exposed condition, and no doubt the feeling of rawness experienced in acute catarrh of the bronchi is due to the cold air acting upon an over-stimulated and exposed mucous membrane. And, further, it can easily be understood that, where this desquamation takes place to an inordinately great extent, the loss of the ciliary action of the columnar cells will seriously interfere with expectoration, and tend to cause the catarrhal products to gravitate downwards towards the smaller bronchi and air-vesicles. This description essentially coincides with what Socoloff found experimentally in animals (*Virchow's Archiv*, vol. lxviii. p. 611), in which he induced an artificial bronchitis by the injection of irritants, such as potassic bichromate, into the air-passages. He states that one of the first changes which ensued was the desquamation of the columnar cells, and that they took no part in the catarrhal inflammatory process.

After the columnar epithelium has been shed, the deeper or germinal layer comes into view, and although this may likewise here and there desquamate, its absence is only of temporary duration and limited to small areas, the denuded portions being apparently very rapidly covered by the neighbouring cells. The whole epithelial covering then assumes quite a different character from that met with in health. So long as the congested state of the parts exists no true columnar cells are produced, and the only attempt at their regeneration is exemplified in the formation of the transitional forms met with in the deeper layers. We then find the bronchial mucous membrane having the appearance represented in Fig. 20, the duration of the attack in the person from whom the bronchus was taken being in all probability about ten days, although I should say, from comparative observations, that the actual part of the mucous membrane represented in the figure was advanced only to the third or fourth day of the catarrh. The same congestion will be seen as in Figs. 18 and 19, but it

ACUTE BRONCHITIS. 31

will now be noticed that it not only affects the inner fibrous coat, but that all the parts of the bronchus exhibit vessels over-distended and engorged with blood. On looking at the

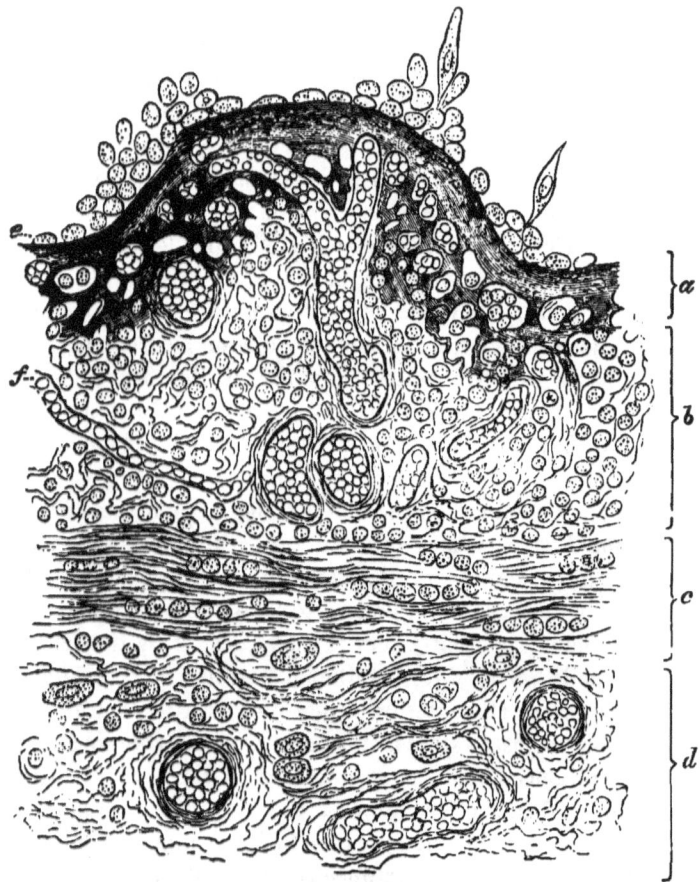

FIG. 20.—Transverse section of entire bronchus. Acute bronchitis: × 450 Diams. *a*, basement membrane; *b*, inner fibrous coat; *c*, muscularis; *d*, outer fibrous coat; *e*, catarrhal cells; *f*, congested vessels of inner fibrous coat.

surface, no vestige of the proper columnar epithelium is visible, its place being taken by the pyriform or *battle-dore-like* cells found in the deep strata of the normal epithelial investment. In no case, however, have I ever found the epithelial covering entirely wanting throughout the whole transverse section of a bronchus, and we shall see that it is this which gives to the secretion its peculiar catarrhal characters. It is not the want of epithelium which is the fault, but rather that the wrong kind is produced, having more or less of an embryonic character. The shape of these abnormal cells is either pyriform, oval, or more or less polygonal, and usually those which lie deepest can be detached in scales from the basement membrane. They are all highly nucleated, and in process of fissiparous division. The flat cells nearest the basement membrane seem to divide first, and very soon the resulting segments rise above its surface (Fig. 18, *e*), being, at first, pyriform, with the slender extremity attached. When this slender extremity has become more and more attenuated it gives way, and they are thrown off into the lumen of the bronchus as catarrhal cells. This process goes on with great rapidity in some parts of the bronchial wall, as seen in Fig. 18, where a great mass of such catarrhal cells has accumulated on the surface. When these catarrhal cells are set free, they become more or less oval or rounded, and very frequently pass into a state of fatty degeneration. They become enveloped in a quantity of the mucus which is constantly being poured out from the orifices of the mucous glands, and together with this constitute what we understand as a catarrhal secretion. The more watery consistence which this has in the early stages of the disease is probably due to an admixture of serous fluid which has escaped from the œdematous mucous membrane, and the yellow colour and tenacious consistence that it assumes at a later period are due to there being less of this serosity and more cellular structures. This abnormal formation of an embryonic epithelium continues so long as the congestion and œdema of the mucous membrane lasts, and the cause of it seems to be the over-stimulation of the epithelial surface, probably from the amount of blood plasma supplied to it. The cellular proliferation goes on so rapidly that there is no time given for elongation and moulding into the complete columnar cell such as we have seen takes place normally; but instead, the divided segments are thrown off in the shape of catarrhal cells as soon as they are formed. Similar processes apparently go on in the mucous glands, especially at their mouths, but not by any means so markedly as on the surface of the mucous membrane itself; they are chiefly concerned, we shall see, in the elaboration of the mucin-holding fluid in which the cells lie. It has been asserted by Rindfleisch and others, not apparently from direct observation, but by analogous comparison with other tissues, that the catarrhal cells are proliferated connective tissue corpuscles, which have escaped on the surface. I have never seen any evidence of this in acute bronchitis, the basement membrane seeming to form an impenetrable barrier to their exit; and even in chronic bronchitis, where this structure becomes much attenuated, I have never seen any direct and clear evidence that they pass through it. There is always a well-marked boundary line between the catarrhal proliferating surface on the one hand, and the infiltrated inner fibrous coat on the other.

While these changes have been going on in the epithelial surface, processes no less striking have been advancing in the substance of the mucous membrane itself. These consist in the dilatation of the vessels, and the infiltration of the inner fibrous coat with cellular structures. The intervals which naturally exist between the fibres of the inner fibrous coat, and which, in silvered preparations, are seen to constitute a rich system of lymphatic vessels and plasma spaces, lined with an endothelium, are now crowded with small round cells, which I shall call leucocytes, irrespective of their origin; and also with cells of a larger size, more irregular shape, and highly nucleated. All these have evidently been in active amœboid motion during life, and run in rows from the basement membrane outwards towards the interspaces between the cartilages and glands, finally making their way into the outer fibrous coat. They are certainly most abundant round the blood-vessels, and, at first sight, one would naturally conclude that they were colourless blood-corpuscles which had exuded. And although there seems little doubt that, as the inflammatory irritation increases, as the vessels become more and more distended, the leucocytes do exude in large numbers, yet it is also quite evident that this is not the only source of this intense cellular infiltration. For if the inner fibrous coat be carefully examined in an early stage of acute bronchitis, the flat endothelial cells and connective-tissue corpuscles which lie in it, especially the former, can be noticed to enlarge, their nuclei, which previously were not particularly evident, have now increased in size, and frequently there are two or more of such in a single cell. The appearance is somewhat like that seen in Fig. 20, *d*, and also in Fig. 21, which, however, are portions of the outer fibrous coat in which the same process is advancing, and in which the flat endothelial cells normally covering the bundles of fibrous tissue, are seen in an active state of endogenous nuclear formation and fissiparous division. The nucleus first becomes prominent and very finely granular, it then elongates and divides into two or more segments, and, co-existently with this, the cell-body becomes hour-glass shaped and similarly divides.

Sometimes three or four nuclei are formed endogenously at one time, and then the cell breaks up into as many separate portions, each portion becoming in its turn rounded and larger, and frequently again undergoing similar endogenous nuclear proliferation and division. The larger highly-nucleated cells seen in the inner fibrous coat, of irregular shape, are undoubtedly formed in this way, while the cell most often noticed, that is to say, a small round cell like a colourless blood-corpuscle, is either the result of this endogenous formation and fissiparous division carried to an extreme degree, or it is simply a colourless blood-corpuscle which has made its way outwards through the coats of a distended blood-vessel.

As a result of these two processes the inner fibrous coat, in from thirty-six to forty-eight hours, begins to show long rows of small round cells, stretching outwards to the intercartilaginous and interglandular spaces, and passing between its fibres into the outer fibrous coat. It is evident that they have been travelling actively, from the various oval and spindle-like shapes which they sometimes assume, and from the fact of their following each other in long rows between the bundles of fibrous tissue. In a short time afterwards, probably in the fourth to the fifth day of an acute catarrhal attack, the cellular infiltration becomes very much greater, and then I find the whole of the lymph spaces entirely choked up by the new cellular products which lie within them. That the direction in which they spread is outwards towards the lung tissue can be easily verified, but that they ever get to the free surface of the bronchial mucous membrane, and are discharged into it, I do not believe to be the case in acute bronchitis. It will be well, however, to examine the evidence on which this statement is founded before going further, and for this purpose, we must look at the basement membrane as it is presented to us from the fourth to the fifth day of an acute catarrhal attack, that is to say, by the time the spaces between the fibres of the inner fibrous coat are distended with cellular structures. We have seen that this basement membrane in health (Fig. 2, *d*) is a delicate, apparently elastic tissue, underlying the epithelial covering, but if we look at it in Figs. 18, *b*, 19, *d*, and 20, *a*, which represent bronchi in an acute catarrhal condition, we should have some difficulty in recognising it as the same structure. Instead of being a thin delicate membrane, it has now become an extremely prominent object, easily seen with a magnifying power of fifty diameters and forming a homogeneous layer which underlies the proliferating epithelium. The reason why it is now so evident is, apparently, that it has become œdematous. Not only, however, does the membrane itself come prominently into view, but its connections with the underlying inner fibrous coat are now rendered very evident. These consist of numbers of branching processes, previously quite invisible, running downwards and losing themselves, apparently by becoming continuous with the fibres of the inner fibrous coat, and evidently forming a continuous attachment of the basement membrane to the latter (Figs. 18 and 20). On examining the basement membrane at this period of a bronchitis, even in this œdematous condition, when it might be thought that any cellular structures which it possessed would have been rendered unusually prominent, I have never been able to make out that any such were contained within it. It appears perfectly homogeneous, unless at its deeper attachments, where it becomes faintly fibrous (Fig. 18). Were it the case that leucocytes made their way through it, and were cast off on the free bronchial surface, they ought certainly to have been visible in this transparent condition of the basement membrane. I have never, however, been able to see these in acute bronchitis, and do not believe that such diapedesis takes place, nor that the cells of the catarrhal secretion are derived from this source. The course which they invariably pursue after they infiltrate the inner fibrous coat is outwards, along the line of the peri-bronchial lymphatic vessels. Its peculiarly homogeneous structure, great thickness, and the fact of there being no interspaces visible in it, all go to favour the view that the basement membrane is the line of separation between what is epithelial and what belongs to the fibrous tissue of the mucosa; and even in the most severe and protracted cases of bronchiectasy this basement membrane is seldom absent, but always maintains its character as a boundary between that which is epithelial and that which is not.

If the attack has lasted for any length of time, the other parts of the bronchi always become the seat of the most marked alterations. The wide interspaces which exist between the bundles of muscular fibres are usually filled with leucocytes (Fig. 20, c), but the muscular fibres themselves do not present any change. In the outer fibrous coat changes co-existent with those in the mucosa are invariably seen after a few days. On referring to Fig. 14 it will be seen that the bundles of wavy fibrous tissue found in the outer fibrous coat, or adventitia, of the larger and smaller bronchi have large connective tissue corpuscles, or endothelia, lying upon them. It is especially in these cells that a sympathetic change takes place when the bronchi are the subject of acute or chronic catarrh, a change apparently analogous to the lymphatic disturbance of distant parts in inflammatory irritation of the skin. The first alteration appreciable in them is that they become unnaturally prominent and finely granular (Fig. 21); the nucleus then enlarges, and frequently contains a vesicular nucleolus. It further becomes hour-glass shaped, the depressions on each side deepening until a complete separation into two parts ensues. This is followed, shortly afterwards, by a similar division in the body of the cell (Fig. 21, *a*), and the resulting portions form two round catarrhal-looking cells, with finely granular protoplasm, and usually several vacuoles contained within them. These further subdivide in a similar manner, and then there are formed either in the adventitious coat, or in the neighbouring processes running off to meet the lobular septa, areas in which large collections of differently-sized actively dividing cells are to be seen. The ultimate result of this proliferation seems to be the production of small round cells like leucocytes, and when the subject of chronic bronchitis is considered the future history of these will be demonstrated, and we shall then perceive that one of the great dangers of an acute bronchitis is this proliferation going on in the neighbouring fibrous attachments. The proliferation is not, however, confined to the immediate vicinity of the bronchi, but soon spreads throughout the lobular septa to the deeper layer of the pleura, the whole of the lymphatic vessels evidently becoming the subject of a catarrh, much the same as that of the bronchi themselves. The changes just described are always most active where there is some congestion, the centre of an actively proliferating area usually containing an artery over-distended with blood-corpuscles, or holding a fibrinous plug in its centre (Fig. 21, *c*).

ACUTE BRONCHITIS.

We have, up till now, considered the changes in the mucous membrane apart from the mucous glands that are contained within it. We have seen that in their structure these glands are to be looked upon as nothing more than processes of the ordinary basement membrane and its epithelial covering, the duct of the gland being lined by both the superficial and deep layers, the fundus being covered by the deepest layer alone, which constitutes the secreting portion.

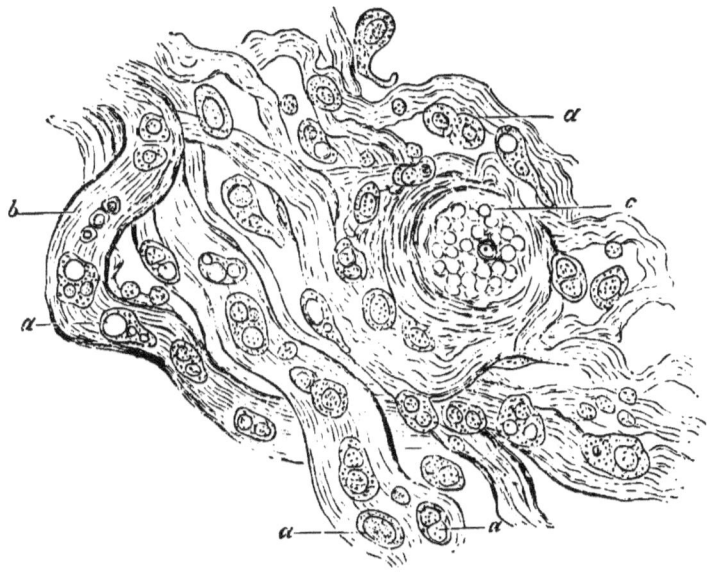

FIG. 21.—Peri-bronchial fibrous tissue. Acute bronchitis—× 450 Diams. *a*, endothelial cells dividing and containing many nuclei and vacuoles; *b*, bundles of fibrous tissue; *c*, congested blood-vessels.

The primary congestion which is noticed in other parts of the mucosa, in acute bronchitis, is usually a very prominent feature in the neighbourhood of the mucous glands, and seems, as in the former, to be the first visible alteration. The next appreciable change seems to be the desquamation of the epithelium lining the mouth and neck of the duct, which is usually seen lying detached in or near the neck, enveloped in a tough plug of mucus. The cells are usually fatty, and in other respects this desquamation is identical with that seen on the free surface of the mucosa itself. The deeper layers of the epithelium, however, are left adherent to the basement membrane, and are seen to be actively proliferating. Similar desquamation seems to ensue in the deeper portions of some of the glands, the desquamated cells becoming very cloudy and granular (Fig. 22, *b*), but in the majority of them an entirely different appearance is presented.

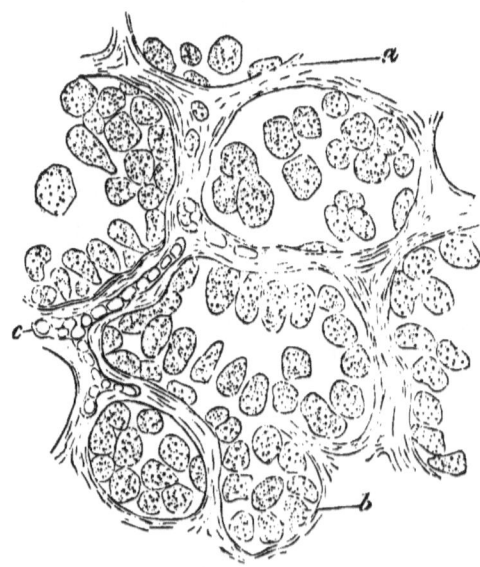

Fig. 22.—Section of a mucous gland. Acute bronchitis— × 450 Diams. *a*, stroma of gland; *b*, secreting cells desquamating; *c*, congested blood-vessels.

This is shown in Fig. 23. It seems to be much the same as Heidenhain has described in the mucus-forming cells of the salivary glands. It can be seen that the whole of the epithelial cells lining the alveoli of the gland are perfectly transparent, their somewhat dark outlines merely sufficing to show the borders of the cells. The cell is also swollen, and its shape altered, while no nucleus is usually to be seen in it. The cause of this is undoubtedly the accumulation of mucus

ACUTE BRONCHITIS.

within the cell, from excessive secretion, and all stages of infiltration and consequent transparency can be noticed, from where, at first, there is a mere clearing up of the natural granularity of the cell at one spot, to where it comes to have the appearance represented in the figure. A few catarrhal

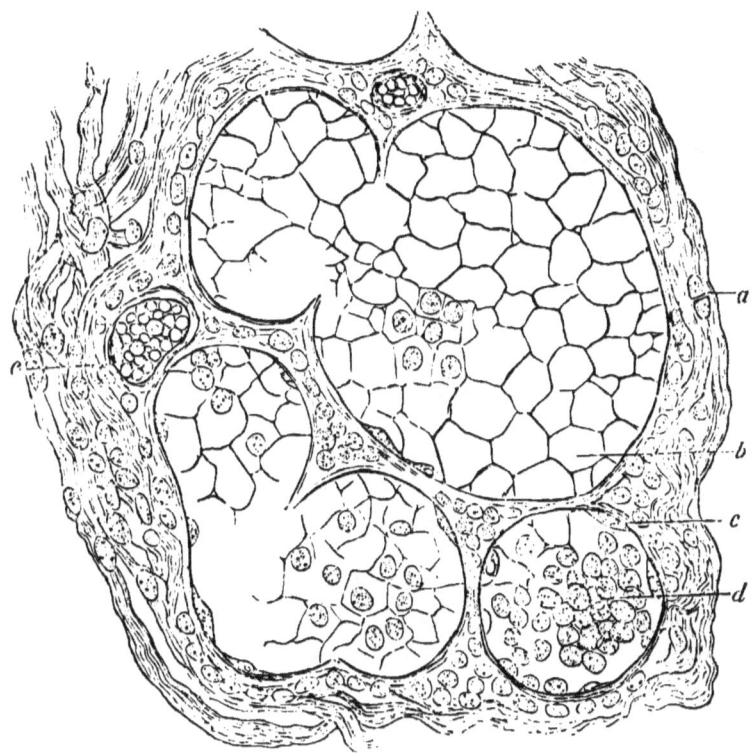

Fig. 23.—Section of a mucous gland. Acute bronchitis—× 450 Diams. *a*, cells dividing in the stroma; *b*, secreting cells distended with mucus; *c*, flat cell layer of epithelium underlying the secreting cells; *d*, catarrhal cells lying in the cavity of the glandular alveolus.

cells are noticed on the surface, but the whole of the appearances seen in connection with the gland go to indicate that the part they take in the formation of the catarrhal discharge is the secretion of mucus, and that although they certainly assist in the production of the catarrhal cells, they are not the source from which the bulk of them is derived. These, which are formed from the epithelial covering of the mucous membrane, merely become mixed with the secreted catarrhal mucus, and the combination of the two gives rise to what we understand as a yellow muco-purulent fluid. The connective tissue corpuscles in the stroma of the gland are observed in an active state of division, and long rows of leucocytes can be seen wandering between the bundles of fibrous tissue (Fig. 23).

The lymphatic glands at the root of the lung are invariably enlarged, and on minute examination the endothelium-like cells lining the stroma are seen actively dividing. The whole of the lymph paths are choked with catarrhal-like cells, similar to those seen in the lobular septa, and the number of lymph corpuscles seen in the gland is increased to an inordinate degree. The vessels are all deeply congested, and, in one or two instances, notably those of children, commencing caseation was noticed at the centre.

Hæmorrhages are also very frequent, and were found in the outer fibrous coat and in the basement membrane in almost every case of acute bronchitis I examined. In some instances they were of considerable size, and, when they occurred in the latter situation, they produced a solution of continuity, with discharge of blood, into the bronchus.

The state of the ganglia and nerve trunks was very interesting both in acute and chronic catarrh. The nerve ganglia of the lung are abundant round the large bronchi at the root, and were originally described by Remak, and more lately they were the subject of a paper at the British Association for 1876, by Dr. W. Stirling. Their function has never been clearly indicated, although judging by analogy, one would suppose that, like the similar structures found in salivary and other glands, and in the stomach, they are concerned in secretion. The only other function that might be attributed to them is in the regulation of the muscular fibres of the bronchi or those in the walls of the bronchial arteries; for, notwithstanding that the pulmonary artery does not seem to be subject to the same laws of vaso-motor action as other arteries throughout the body, there is no reason to believe that the bronchial arteries are to be included in the same category. The fact of their being so numerous looks, however, as if they had surely some other function than this, and the appearance they have in acute bronchitis, when the glands of the mucous membrane are abnormally active, would favour the idea of their being concerned in the secretion of

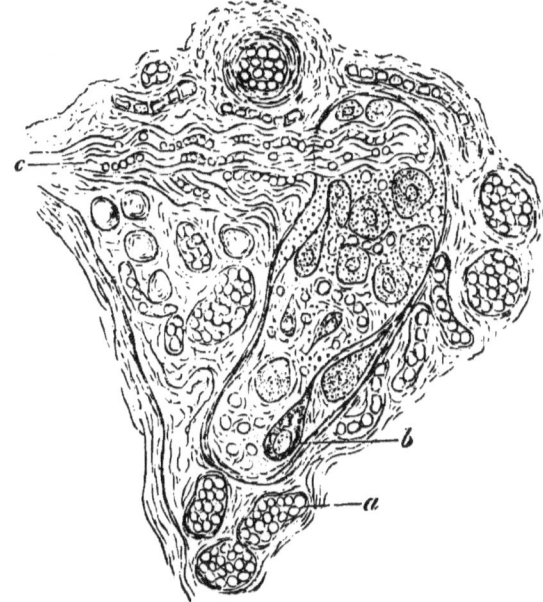

Fig. 24.—Section of a bronchial ganglion. Acute bronchitis— × 450 Diams. *a*, congested blood-vessels around ganglion; *b*, ganglion cells; *c*, congested capillaries of nerve trunk.

mucus. A representation is given of one of them in Fig. 24, taken from a child who suffered from acute bronchitis after whooping cough; and although it might be urged that the abnormal amount of congestion seen in it and its neighbourhood might have had something to do with the essentially spasmodic nature of the disease, yet this seems discredited by the fact of its being equally great in other cases either arising idiopathically or as sequelæ to certain febrile conditions. The most intense congestion of the surrounding vessels is always noticed, and the minute capillaries which run into the ganglion and nerve trunks are all engorged with blood. I cannot say, of course, that the congestion of the ganglia and the hypersecretion of mucus were cause and effect, but I would merely suggest it as probable that the over-excitement of these nerve-centres may have had something to do with it. Had they been concerned in vaso-motor action it might have been expected that their over-stimulation would have given rise to spasmodic contraction of the muscular coat of the *bronchial* arteries. Such, however, was not the case, one of the most marked phenomena being, as previously indicated, dilatation and evident exhaustion or paralysis of the muscular coat. The excitement of mucous secretion from the bronchial, tracheal, and nasal mucous membranes, on the stimulation caused by mere sudden variations in temperature, would point also to its being in all probability a reflex phenomenon, and it seems only reasonable to suppose, as the ganglia are so intimately associated with those bronchi in which mucous glands are found, that they are possibly the immediate centres for that reflex act.

The first indication of recovery taking place in acute bronchitis, seems to be the diminution of the congestion of the mucous membrane. The vessels seem to recover their wonted tone, and the circulation goes on more freely. Simultaneously with this, the excitement of the epithelial surface seems to diminish, the proliferation of the epithelial cells becomes less active, and, instead of being thrown off in an embryonic condition, as a muco-purulent secretion, they are again converted into fully-developed columnar cells. The process by which this is accomplished is identically that described in their formation, so that, little by little, the epithelial covering assumes its normal appearance. The cellular infiltration of the wall of the bronchus is got rid of apparently by means of the lymphatics, but, in cases which become chronic, this absorption causes great surrounding irritation, ending in the formation of more or less interstitial pneumonia.

Chronic Bronchitis.

The etiology of chronic bronchial catarrh is so varied that I have thought it best to arrange the subject according to the agents which are instrumental in its production, trusting that my observations on the different forms of the disease may thus be rendered more intelligible. I shall, at present, chiefly examine the condition of the bronchi alone, leaving the description of the pulmonary complications, which almost invariably arise in connection with chronic catarrh of their mucous membrane, until afterwards. The divisions that I propose making in the treatment of the subject of chronic bronchitis are according as it is due to one or other of the following four causes:

(*a*.) An acute attack.
(*b*.) Valvular lesion of the heart.
(*c*.) Inhalation of foreign matters.
(*d*.) Chronic interstitial nephritis.

Chronic Bronchitis following an Acute Attack.

When an attack of acute bronchitis has reached its maximum intensity, one of three things may happen if the patient survives. The catarrhal inflammation may undergo resolution, or it may become complicated with catarrhal pneumonia; or, finally, it may become chronic. The first of these results is usually what follows in persons of robust constitution and in middle life; the addition of a catarrhal pneumonia is, of course, a very serious complication; while the production of a *chronic* catarrhal state of the bronchial mucous membrane, although perhaps not fraught with immediate danger, is the most persistent, and probably the least amenable to treatment. There is little doubt that most of the cases of chronic bronchitis have originated in an acute attack which has never undergone complete resolution, but has left the bronchial wall in a more or less impaired condition, so that the slightest undue exposure is sufficient to cause a great exacerbation of the symptoms. Sooner or later pulmonary and cardiac complications arise, which, taken along with the blocking-up of the air-passages with catarrhal fluid, prove too much for the patient's strength, and give rise to a fatal result. I shall endeavour to show what takes place in the bronchi and their surroundings in such cases, and what the mechanism is which prevents a favourable result being speedily brought about.

The lungs are usually extremely emphysematous, and cover the greater part of the heart in front: their lobules are particularly evident from over-distension, and from the deposit of black pigment lying between them. When the lungs are removed from the chest and cut into, partial or complete collapse takes place. The mucous membrane of the bronchi is much congested and of a deep bluish-red colour; that of the lower part of the trachea is also intensely congested, and of the same cyanotic tint. The whole of the smaller and middle-sized bronchi are filled with yellow muco-purulent discharge, which can be squeezed out in the form of little pellets. This is sometimes extremely viscid, of a greyish colour, of mucous or jelly-like consistence, and adheres firmly to the bronchial wall. The mucous membrane is perfectly smooth and shining: no deficiencies are apparent in it with the exception of the mouths of the mucous glands, which are seen as little pin-point depressions when light is allowed to fall obliquely upon them. There is no evidence of the surface being in a granulating condition. When looked at a little more closely it can be noticed that the smooth glistening appearance of the mucosa is due to the basement membrane which covers it. If this be touched with a sharp-pointed instrument it will move freely over the inner fibrous coat of the mucous membrane on which it rests. It can, further, be dissected off with a sharp scalpel, and if it be then placed in a neutral fluid, and examined microscopically, it will be found to have the following characters: On its free surface lies the epithelium, or, at any rate, the embryonic forms of it met with in chronic bronchitis, while underlying it are the longitudinal bundles of yellow elastic fibres. It is perfectly homogeneous, has not any fenestræ or other apertures, and does not exhibit any cellular structure. If the cover-glass be pressed down so as to endeavour to destroy it, this will be accomplished with difficulty, for instead of breaking into pieces, it will be found to stretch readily, and to retract when the pressure is removed. When treated with glacial acetic acid, no change seems to follow, in this respect resembling the elastic fibres which underlie it. The epithelium, which can be scraped off from its surface, shows comparatively few columnar cells, the usual forms met with being round or "transitional" in character.

The whole mucous membrane is much thickened and thrown into longitudinal folds, and the elastic fibres of the inner fibrous coat can be noticed as greyish-coloured bands running longitudinally. The smaller bronchi are usually dilated, but this depends upon certain circumstances to be afterwards described, the dilatation being by no means a necessary feature. In certain instances the lumen of the larger bronchi seems to be narrowed, from the great thickness of the mucous membrane. The adventitious coat of the bronchi, even where there is no marked interstitial pneumonia, is also usually thickened. The bronchial glands are invariably enlarged and pigmented.

In Figure 25 is represented such a bronchus as that just described, transversely cut, after having been specially prepared for microscopic purposes. It was of medium size, and is magnified fifty diameters, and reduced one half. It will be observed, in the first place, that the wall is greatly thickened, and that the mucous membrane is thrown into folds. There is no proper columnar epithelium on the surface, but, in place of this, numbers of small bud-like projections (a) are seen on the basement membrane. The thickening in the bronchial wall is almost entirely due to cellular infiltration, which is to be seen not only in the mucous membrane, but extends also to the outer fibrous

CHRONIC BRONCHITIS.

coat. The muscularis has nearly all disappeared, unless at one spot (*b*), where a few muscular fibres are still left. The glands and cartilages have entirely vanished, their place being taken by dense cellular infiltration. The cause of the irregularity of the free surface in this case seems to be that the cellular infiltration of the mucosa is greater at one part than another; but it will be noticed that in no instance

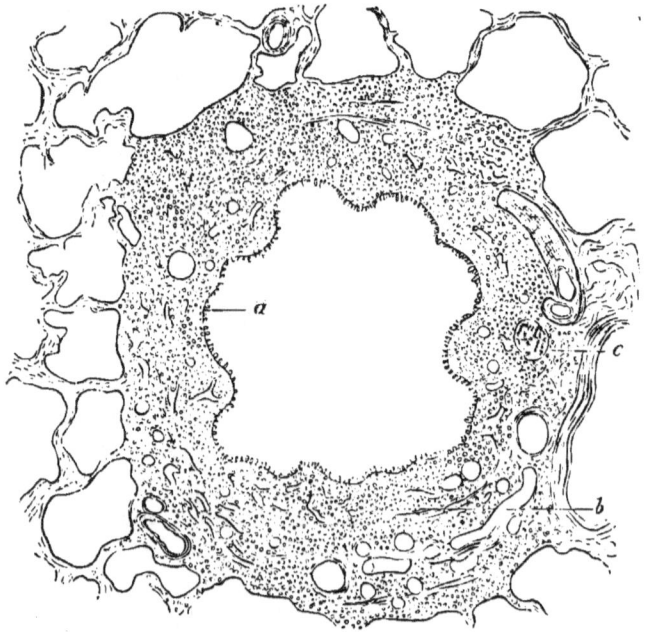

FIG. 25.—Trans. sect. of a bronchus in a state of chronic catarrh. × 50 Diams. (reduced one half). *a*, germinating epithelium, placed on basement membrane; *b*, remains of muscularis; *c*, a small nerve.

has the smooth outline given to it by the basement membrane broken down. There is a perfectly clean-cut margin, without any evidence of a granulating surface. The mucous membrane of the same bronchus is represented more highly magnified in Fig. 26, and it will give an idea of the average condition of such bronchi. The true character of the bud-like process, seen with the lower magnifying power, can now be made

48 CHRONIC BRONCHITIS.

out. They are the pyriform and rounded catarrhal cells, which are being thrown off from the deepest or flat cell layer of the epithelial strata, in the same way as they are cast off in acute bronchitis. Some of the flat cells are seen partially raised from the basement membrane at c, and the

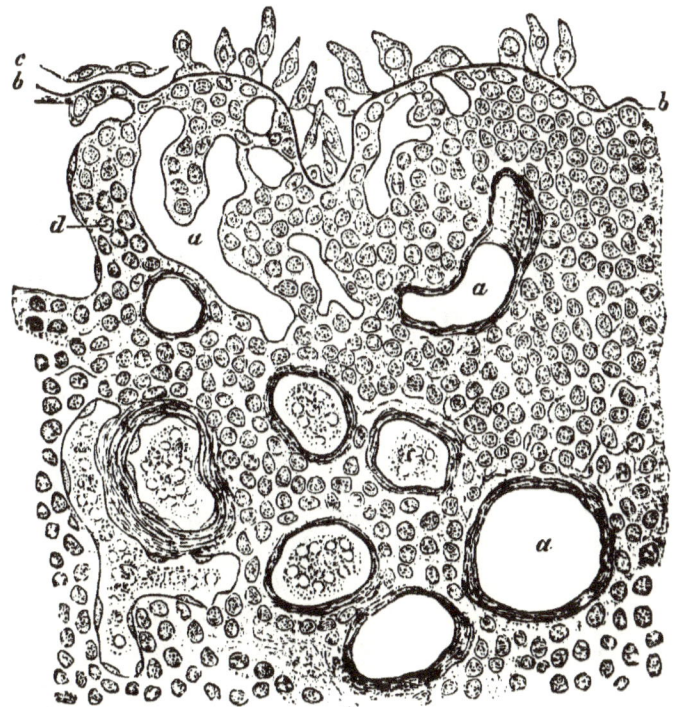

FIG. 26.—Mucosa of bronchus seen in Fig. 25. ×450 Diams. a, dilated blood-vessels; b, basement membrane; c, desquamating epithelium; d, small cell infiltration of mucosa.

different stages in the formation of catarrhal cells, from the time when there is a mere oval elevation to where this becomes attenuated and finally detached, are also represented. No fully-formed columnar epithelium was found in any part of this bronchus, the whole of the epithelial cells being thrown off in an embryonic condition.

CHRONIC BRONCHITIS.

Immediately underlying the layer of germinating epithelium a dark line is noticed in the figure on which the epithelium rests. It is *the elastic basement membrane* described above, which, in this instance, was not particularly œdematous, although this is by no means invariably the case. Its underlying attachments cannot be seen, probably on account of the cellular infiltration, but it will be observed to form a continuous layer on the surface and to give the smooth character to the mucosa. Distended capillaries projected into it, and every here and there small ruptures of these, with hæmorrhage into the bronchus, had occurred. I have always found this basement membrane present in chronic bronchitis, and it usually forms one of the best-marked objects. It appears, in some cases, to be much thickened, and to take the place, to a great extent, of the epithelial covering. The question again arises, as in acute catarrh, whether there are any openings or fenestræ naturally present in it, and whether cellular structures are capable of passing through it from below. In chronic bronchitis the underlying inner fibrous coat has its interstitial spaces over-distended with cellular structures, which lie close to the attached surface of the basement membrane, and it becomes a very difficult matter to decide whether they pierce through it and project on the surface. There is a fallacy connected with this observation which would be certain to deceive one at first sight. It is very seldom that the basement membrane can be cut in an exactly transverse direction; the smallest deviation in the perpendicular position of the bronchus, in cutting it, naturally gives a more or less oblique view, so that it cannot all be seen at the same level. It accordingly happens that at one time we see certain cellular structures on the free surface, projecting into the bronchus and undoubtedly having no further attachment than to the basement membrane itself: while at another, by a slight alteration of the focus, several cells corresponding in appearance may be seen at a deeper level, which look as if their attached extremities passed through the basement membrane into

E

the inner fibrous coat, on account of the oblique direction in which it is lying. From the most careful examination of this point in many instances of chronic bronchitis, either where the basement membrane has been œdematous, or where it has had its natural appearance, I am inclined to believe that in this, as in acute bronchitis, nothing of a cellular nature ever gets from below on to the free surface. It seems to form an impenetrable barrier to the exit of leucocytes or other cellular structures, and although it is frequently thrown into folds, from the irregular accumulation of cellular structures under it, yet I have never been persuaded that any of these make their way through it. It maintains its invariably homogeneous aspect, without there being a single break in its continuity. This is evident on examining its relations in Fig. 26; for although there was exceptionally little œdema in this instance, yet none of the underlying cells appear to have escaped through it.

It will, therefore, be apparent that this is a structure of the greatest importance, for, if it be correct that cellular bodies cannot pass through it from below, it follows that unless the enormous accumulation of such, which takes place in the mucosa in chronic bronchitis (Fig. 25), degenerates, or is got rid of in some other way, the only means of exit, the only direction in which the protoplasmic structures can travel, is outwards into the lung tissue. As the accumulation increases, and, consequently, as the pressure from it becomes more marked, we can see that the cells, which lie in the plasmatic spaces of the mucosa, all tend to run outwards; they follow the direction of the lymph stream, and they are very soon found infiltrating the peribronchial and periarterial fibrous tissue. There is not the slightest doubt that the presence of the basement membrane is a most important determining cause in the direction which they take, for, if it were not so, there is no reason to believe, after the epithelium is partially shed, that they could not, and would not, wander in towards the free surface of the bronchus, and be thrown off from it as they are from a granulating wound. They would naturally take this direction, being that of least resistance; but, instead of this, they seem to accumulate in the fibrillar interspaces of the inner fibrous coat, until, from their pressure, they distend these sufficiently to permit of their passage outwards into the outer fibrous coat and lobular septa. Their further course will afterwards be traced, and it will be shown that many of the disastrous results ensuing from chronic bronchitis are due to the effect they produce on neighbouring parts.

It would, therefore, seem as if the presence of this elastic, apparently indestructible, basement layer was in reality the great cause of an acute attack of bronchitis becoming chronic; for if the cellular accumulation in the mucosa, in an acute attack, could easily escape on to its surface, there is every reason to believe that the vessels would regain their wonted tone, and that the epithelium would resume its normal appearance. Indeed, the conviction is irresistible in the examination of such bronchi that if the basement membrane could be broken down or scarified, the cellular pressure on the small blood-vessels would be relieved, and the part would have a chance of resuming its natural functions; or, even, if degeneration of the inflammatory products could be brought about, that the same result would be gained. Such degeneration unfortunately never seems to occur in chronic bronchitis, and the only course evidently left for the inflammatory cells, which have distended the mucosa, to pursue, is to wander outwards, and to get into the surrounding interstitial tissue of the lung itself. The effect of the basement membrane seems to be very much the same as that of a fascia, in determining the direction that the pus in an abscess will pursue.

The portion of Fig. 26 below the basement membrane corresponds to the inner fibrous coat. It represents a bronchus of about the same size as that shown in Fig. 2, and, if the relative thickness of the inner fibrous coat in each be compared, an idea may be formed of the amount of cellular accumulation which has taken place. This could hardly be recognised as the same structure as that represented in Fig. 2 at *c*.

The enlargement has been produced mainly by the cellular accumulation in its fibrous interspaces, but partly also by the great distension of its blood, and probably also lymphatic, vessels. At the same time its fibrous tissue has disappeared, or its interspaces have become so distended that its fibres can be seen only here and there. The appearance of the inner fibrous coat is more that of an abscess, only that none of the cellular structures within it are becoming fatty. All of them, judging from the varying shapes which they assume, and from the manner in which they adapt themselves to the irregularities of the part, seem to be in an active state of amœboid movement. When traced further outwards they are seen running in lines towards the spaces between the cartilages and glands, and they can be noticed subsequently infiltrating the outer fibrous coat. They are usually about the size of a blood leucocyte, but some of them are a little larger. Their source seems to be the same as in acute bronchitis.

The muscular coat of the small arteries is usually much thickened (Fig. 26), and the capillary vessels, from being channels capable of allowing a single blood corpuscle to pass, are distended into large sinuous spaces. The nuclei of the endothelium in all the blood-vessels are unusually prominent, so that they can be easily noticed on transverse section. There are also numbers of other vessels whose walls seem to be constituted by a single endothelial layer, and which frequently envelope a small artery. They contain a finely granular mass, with cells occasionally in it, and seem to be over-distended lymphatics. They are met with in considerable abundance close to the cartilages.

We saw that in acute bronchitis *the muscular coat of the bronchus* is not appreciably altered further than by the cellular infiltration which takes place into the spaces between its fibres. When the disease becomes chronic, however, changes of a much more decided character are to be seen in it. Its function seems in great part to be to prevent over-distension of the bronchus in forcible expiratory efforts—a regulating action very much like that of the muscular coat in a small artery. When an artery passes into an organ which is in a state of cirrhosis its smaller branches become much compressed, from the contraction of the newly-formed cicatricial tissue which lies around them. Their lumina are narrowed and the blood has naturally great difficulty in passing through them. Under such circumstances, an increased amount of blood-pressure is thrown upon the main branches, behind the points of obstruction, and no doubt a similar reaction takes place throughout the whole arterial system to a certain extent. I have invariably found that the muscular coat of such arteries, no matter what the organ may be, is more or less hypertrophied, usually to a very great extent. The lumen of the vessel, at the same time, is *not dilated*, but generally somewhat narrower than in vessels of corresponding size when normal. The explanation that I would give of the production of the muscular hypertrophy in such a case is simply that it is a provision in the artery to resist over-distension, and to prevent the loss of its proper tonicity. There is an undue amount of pressure brought to bear upon the muscular wall from impeded circulation, and were some provision of this kind not made, the lumen of the vessel would no doubt become unduly enlarged. This, however, does not take place; it remains of its usual size, or is even smaller than normal. In confirmation of this theory of the cause of the hypertrophy of the muscular coat of the artery, we may compare it with what occurs in vessels unprovided with a muscular coat under similar circumstances. If the lung, liver, and kidneys be examined in regurgitant mitral disease the whole of their capillary system of blood-vessels will be found to be immensely dilated, and their coats attenuated. The same thing would undoubtedly happen in the arteries were there not sufficient muscular hypertrophy to counteract the regurgitant pressure, and to maintain the tone of the vessel.

It has always seemed to me that a comparison can be drawn between the muscular coat of the artery and that of the bronchus under such analogous conditions. The forcible expiratory efforts made by those suffering from chronic bronchitis in coughing must bring a very considerable amount of elastic pressure to bear upon the whole of the respiratory passages, from the glottis downwards, and this pressure being, when the glottis is closed, equal throughout, will act most deleteriously on the weakest part. We would naturally expect that the air-vesicles would suffer most, and the almost constant over-distension of them in chronic bronchitis entirely bears this out. In the human lung they do not seem to contain many muscular fibres in their wall, and having only their natural elasticity to resist the inordinate pressure brought to bear upon them, they dilate most easily of all the respiratory passages. The bronchi, in many cases of vesicular emphysema accompanied by chronic bronchitis, are much dilated, no doubt from the same cause, but always to a less extent than the air-vesicles. In similar instances of vesicular emphysema, however, there is sometimes no co-existent dilatation to be noticed in the bronchi; indeed, the lumen often seems to be unnaturally small; and, if a microscopic examination of the muscular coat be made in such a case, I believe that the cause of the non-distension will be found in its hypertrophy. From being a thin delicate lamina of muscular fibres it becomes a layer of thick coarse bundles; and, judging from the manner in which the mucous membrane is frequently thrown into folds in such a case, it has evidently been in a state of tonic contraction. It would, therefore, seem that here the muscular coat of the bronchus plays a part similar to that of the muscular coat of an artery when there is a tendency to over-distension.

What, however, is the state of the muscular coat where bronchiectasy and vesicular emphysema co-exist, and where there is no ulterior cause for the bronchiectasy, such as chronic interstitial pneumonia, or purulent accumulation within the bronchus? A good example of it can be seen in Fig. 25, where it will be noticed that it is only here and there that any trace of the muscular coat can be observed. The greater part of it has been supplanted by cellular effusion, and the fibrous tissue of the mucosa has also been mostly removed, its place being taken by inflammatory cellular products. Is it to be wondered at, that such a bronchus, even although the bronchial wall is much thickened, should be less able to withstand forced expiratory efforts? It is more matter of surprise that greater dilatation does not take place than usually happens, the explanation being apparently that the chest-wall prevents it. The manner in which the muscular fibres are removed seems to be by simple atrophy. The pressure of the cellular accumulation is so great, that even although the blood supply is abundant, the muscular fibres become resolved into granules, and are ultimately absorbed. Pressure is, no doubt, the direct cause of their atrophy; and although it seems to be mainly due to the cellular accumulation in the mucosa, yet the distended vessels which surround the muscular coat on all sides must materially aid in bringing about this disastrous result. Were the cellular effusion in the bronchial wall to organize into fibrous tissue, an artificial support might be afforded, and bronchiectasy prevented. I do not remember, however, having noticed any fibrous organization, in a case of simple chronic bronchitis uncomplicated with interstitial pneumonia. The inflammatory effusion seems to remain in its embryonic condition; the cells merely wander out into the peribronchial lymphatics as soon as the mucosa becomes over-distended, and fresh accessions are constantly made to replace those which have been thus removed.

The same small cell infiltration is also to be seen in the outer fibrous coat, the fibrous tissue of which, in some instances (Fig. 25), is rendered invisible, either from atrophy or from the distension of its fibrillar spaces with cellular structures. The cellular infiltration is frequently continued into the perivascular fibrous tissue; but here, in addition, the large endothelial cells covering the bundles of fibres are to be seen in a state of germination. The appearance which these endothelial cells in the perivascular fibrous tissue present is very much like that in the peribronchial fibrous tissue in acute bronchitis (Fig. 21), only much more widely distributed. The first deviation that I have noticed is that they enlarge, the body of the cell becomes more evident than it is normally, and they seem to be raised from the surface of the fibrous bundle on which they lie. The nucleus also begins to participate in the enlargement, soon becomes hour-glass shaped, and subsequently divides, so that most of these large granular cells usually appear to have two or more nuclei. Some of them contain, in addition, a large vacuole, and it very often seems as if the nucleus had escaped from the protoplasm, and left a vacuole in its place. In other instances the protoplasm appears to be destroyed, and the nuclei to be set free. In no case, however, are the nuclei destroyed, the ultimate result of the cell proliferation being the production of large depots of free nuclei. These are more numerous in some parts than in others, and, usually, where the cellular infiltration is not so extreme as to obliterate the surroundings, loose bands of fibrous tissue are seen running throughout each deposit. When the surroundings of such a small cell collection are carefully examined, it can be noticed that long lines of free nuclei, or, as they might now be termed, small round cells, are continued between the fibrillar spaces of the lobular septa communicating with the bronchial and arterial sheaths. The above described cellular infiltration of the bronchial and arterial surroundings usually extends through out the whole lung tissue, being, frequently, as well marked near the pleura as round the larger bronchi and arteries.

The cartilages in the bronchial wall, in chronic bronchitis, are more or less altered, the commonest lesion being atrophy and absorption, apparently from pressure; and very frequently there are not any cartilages to be seen, or only the smallest remains of them, in bronchi which otherwise should have possessed them. The vessels around them are usually abundant, so that their destruction cannot be dependent on mal-nutrition. It seems to be more the result of direct pressure, the same as that which causes the removal of the muscular coat. The atrophy is always best seen where the cellular accumulation in the bronchial wall is greatest, or

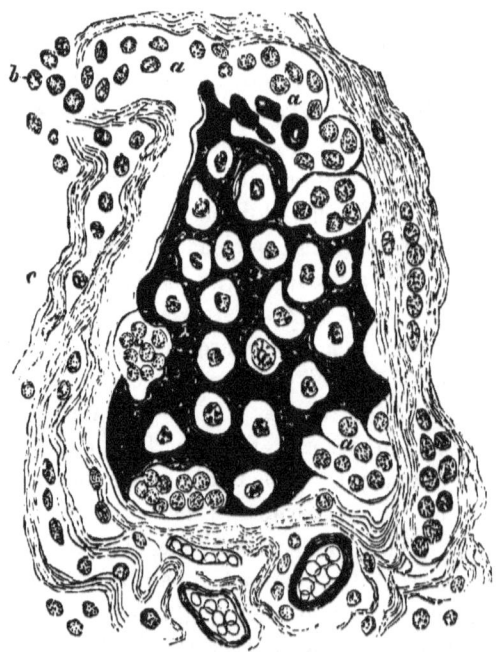

FIG. 27.—Bronchial cartilage undergoing atrophy. × 450 Diams. *a*, large spaces at periphery filled with cells; *b*, the same cells apparently wandering into the spaces; *c*, surrounding fibrous stroma.

where, in addition, the bronchitis is complicated with interstitial pneumonia.

The appearance they have is represented in Fig. 27. The perichondrium has disappeared, and its place is taken by cellular deposit, or, at least, the cellular deposit has so distended its interfibrillar spaces that it can no longer be recognized. A somewhat wide interval lies around the remains of the cartilage, filled with small round cells. The cartilage itself is misshapen and much reduced in size. Instead of being a smooth bordered ellipse, the continuity of its outline is every now and again interrupted by some irregularity, due to the absorption of the hyaline matrix, forming large spaces (Fig. 27, *a*), in which numbers of small round cells are noticed. These cells evidently pass into the absorption spaces from without, as they can be traced in lines continuous with the surrounding cellular infiltration. Portions of detached matrix can also be noticed lying in the cavities, having evidently become separated from the parent mass. What appeared to be newly-formed blood-vessels were also seen running into these cavities, and blood-corpuscles could sometimes be noticed lying in them, apparently resulting from small hæmorrhages. Simultaneously with the occurrence of these changes in the matrix an equivalent process of destruction is noticed in the cartilage cells. Some of them have disappeared, but others have become shrunken or very granular, and frequently one or two oil-globules can be seen in or around them. I have never observed them proliferating, the whole process seeming to be one of atrophy and substitution of extraneous embryonic cellular products in their place. Much the same description of the process has been given by Filz as occurring in the cartilages of bronchi the subjects of bronchiectasy; and when we consider the structure of bronchiectatic cavities it will be shown that this process of absorption of the cartilages is one of its best-marked features. There cannot be the slightest doubt that the removal of the cartilages must materially aid in the formation of certain kinds of bronchiectatic cavities; and, further, their place being occupied by embryonic cellular products, will tend to favour dilatation.

The mucous glands, in chronic bronchitis following an acute attack, are found in varying degrees of abnormality. They are never in a natural state, and, in some of them, the whole secreting parenchyma and acini seem to be destroyed. This is the result of extreme cellular infiltration of the stroma of the gland. In Fig. 23 a certain amount of this cellular infiltration of the stroma is represented as occurring in acute bronchitis; but in chronic bronchitis following an acute attack, little of the stroma is to be seen, its place having been taken by dense cellular infiltration. Towards the neck of the gland this sometimes amounts to the formation of an abscess, which apparently discharges its contents into the trumpet-shaped opening. Such an abscess is represented in Fig. 28. It was placed close to the mouth of a mucous gland, and was about to open into it. It had pushed all the neighbouring tissues aside, and was beginning to soften in the centre.

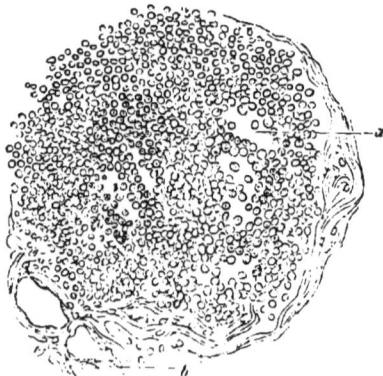

FIG. 28.—Small abscess at mouth of mucous gland. *a*, abscess cavity; *b*, a neighbouring blood-vessel. × 300 Diams.

It is rarely that an acinus is completely invested by epithelium. The secreting cells usually present more or less of the appearance represented in Fig. 23, where they are distended with mucus. Their disintegration is also commonly seen. The gland-cells become partially transparent from mucous distension, rupture, and are then resolved into granular debris, their place being taken by new cells, evidently formed from the deeper layer. In certain instances, however, the cellular infiltration of the gland-stroma completely destroys the acini, and the whole gland may then be removed and supplanted by embryonic cellular tissue. The ducts and mouths of the mucous glands are usually much dilated, and the cause of the dilatation is, apparently, the accumulation of mucus within them. The epithelium lining the ducts is seldom normal in appearance, the cells being usually more or less transitional in shape.

Atheromatous disease of the middle-sized branches of the pulmonary artery, in chronic bronchitis, is by no means uncommon. I have seen it associated with fatty metamorphosis of the muscular fibres of the heart; but whether the two have any mutual relationship I cannot say.

Chronic Bronchitis following Valvular Lesion of the Heart.

The interference which any valvular lesion of the heart produces in the circulation of the blood, gives rise to more or less regurgitant pressure in every organ. Mitral regurgitation will naturally be that which will more directly affect the lungs, but any valvular incompetency sooner or later comes to cause some amount of obstruction to the blood circulating through them, and induces certain well-marked bronchial and pulmonary symptoms. The subjects of such valvular lesions are constantly liable to attacks of bronchitis accompanied by slight hæmoptysis, and I shall now proceed to examine the condition of the bronchi in such cases, and to indicate what the cause of the bronchial catarrhal attacks is.

When the bronchi are laid open, the most striking feature is the intense congestion and deep cyanotic tint of the mucous membrane. It is covered with more or less mucus, and when the smaller bronchi are squeezed, a quantity of frothy serous fluid and mucus exudes.

The lung substance is also intensely congested, and numbers of punctiform hæmorrhages can be noticed between the superficial and deep layers of the pleura. All throughout the lung there are ill-defined indurated patches, of a brownish-red colour, the so-called "brown induration." If these be squeezed, they will be found to contain much less serous fluid than the surrounding lung-substance, and to be hard, comparatively non-vesicular, and somewhat dry. Their border is not at all well marked. It seems to lose itself insensibly in the surrounding lung substance. At the same time there are frequently several wedge-shaped hæmorrhagic infarctions towards the periphery of the lung. They are usually red in colour, becoming yellow as they grow older. Their base is placed towards the pleura; the border is extremely sharply defined; they are hard and non-vesicular, and sink in water. Pleurisy is generally present to a small extent, but inflammatory affections of the pleura are by no means characteristic of this lesion of the lungs.

The appearance which the bronchi have in such a case is shown in Fig. 29. The lumen of the bronchus may be narrowed or dilated, according to the amount of expansile pressure which has been exerted upon it and its ability or inability to withstand this. The mucous membrane is thrown into folds, apparently from the great turgescence of its capillary blood-vessels. There is not usually any epithelium on its free surface, or if there is, it merely consists of small germinal buds of a transitional character. The basement membrane is thick and œdematous, and seems to take the place of the epithelium in forming a protective covering. The chief feature, however, is the extraordinary distension of the capillaries (b) in the bronchial wall. Every minute branch is dilated into a vessel quite easily seen with a magnifying power of fifty diameters. They are all filled with blood, which does not readily leave them when the part is cut across; and they are tortuous and varicose, pushing the basement membrane forwards into irregular folds. Rupture of a superficial branch is common, the basement membrane being torn across, and the blood escaping into the bronchial lumen. The whole of the interfibrillar spaces of the mucous membrane are widened and apparently œdematous, but small cell infiltration into them is not a marked characteristic. In this latter respect the bronchi in

CHRONIC BRONCHITIS.

valvular lesion of the heart present a marked contrast to those affected with chronic bronchitis uncomplicated with cardiac insufficiency, and commencing as an acute attack.

In both cases the bronchial wall is thickened; in the former, however, it is due mainly to capillary distension and œdema of the fibrous tissue, while in the latter it owes its origin in great part to cellular infiltration. This capillary

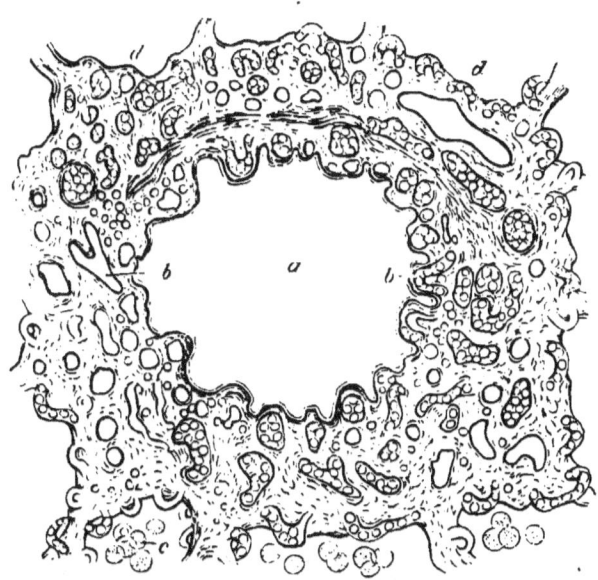

FIG. 29.—Transverse section of a bronchus, regurgitant mitral disease. × 300 Diams. *a*, bronchial lumen, with mucous membrane thrown into folds; *b*, congested capillaries; *c*, desquamated and pigmented alveolar endothelium; *d*, alveolar cavities of lung.

dilatation not only affects the bronchial wall, but extends to that of the alveolar cavities. In Fig. 29 the distended capillaries in the alveolar wall are represented at *d*, and it will be noticed that they are not only distended, but tortuous and congested, and project into the alveolar cavity. Lying within the alveolar cavities there are numbers of large round brown pigmented cells (Fig. 29, *c*), which can frequently be seen in process of desquamation from the alveolar wall.

They undoubtedly represent young and germinal alveolar endothelium, which has been cast off before it has reached maturity; and the pigment which they contain is blood pigment derived from small hæmorrhages and effused blood-serum. Every here and there groups of three or four air-vesicles can be seen filled with blood, and these, along with the desquamated endothelium and capillary distension, give rise to the patches of "brown induration." The mucous glands of the bronchi are usually distended with œdematous fluid and mucus, and their epithelium is more or less in a state of disintegration.

I would therefore say that the cause of the chronic bronchial irritation is a mechanical one, namely, regurgitant blood-pressure, producing a varicose ectasy of the bronchial capillaries, and that the lesion of the bronchi is not a true bronchitis. The surface of the mucous membrane is from its œdematous condition unsuited for the growth of epithelium, which, accordingly, is usually absent, and hence we do not find that there is the same amount of cellular structures in the expectoration as in ordinary bronchitis.

The greater part of the secretion is composed of mucus and serous fluid, and the sources of each of these are respectively the mucous glands and the general surface of the mucous membrane. The cause of the desquamation of the alveolar endothelium is, as has been shown by Buhl (*Lungenentzündung, Tuberkulose, und Schwindsucht*), merely mechanical, and due to the œdema of the alveolar wall and loosening of the endothelial cells. These cells, while yet immature, are thrown off into the alveolar cavity, where they absorb a certain amount of blood-colouring matter and become pigmented. I believe that the lesion in the bronchi is analogous to this, and is not a true bronchitis.

Chronic Bronchitis due to inhalation of foreign matters.

I have had very favourable opportunities of studying the effects of inhaled foreign bodies on the bronchi in the appearances presented in the coal-miner's lung. Bronchitis, quite apart from the disease known as "black spit," is extremely common in coal-miners in middle and old age, and, curiously enough, the secretion they expectorate is yellow, muco-purulent, and not pigmented, even although, after death, the lungs may be found to be perfectly black in colour. There is not the slightest doubt that the greater part of the material inhaled is coal-dust, the particles not only presenting the microscopic structure of coal, but even sometimes the special bands of the particular coal in which the subject of the disease may have worked. Part of it, however, seems to be carbon, especially in the lungs of old coal-miners, who have worked in mines when lamps which gave off a great deal of smoke were used, and when ventilation was not so completely carried out as it is at present. In Fig. 31 some of the commonest shapes of the inhaled particles are seen. The greater number of them are either small granule-like bodies, or they are little acicular spicules, and one can easily see how their lancinating edges might cut into a texture. I have examined the bronchi in many such cases, and have invariably found the same appearances. One of the most remarkable points revealed by microscopic examination is that the larger, middle-sized, and most of the smallest bronchi are *not pigmented*. The catarrhal discharge which is found lying within them is of the usual yellow muco-purulent character, and neither to the naked eye, nor microscopically, is there to be found any evidence of the inhaled particles making their way through the mucous membrane of the bronchi. The whole of it seems to gain admission to the lung through the air-vesicles, infundibula, and some of the smallest sized bronchi. In

CHRONIC BRONCHITIS. 65

Fig. 30, the appearance of a bronchus, taken from a coal-miner who died from bronchitis, is represented as seen when examined with a low magnifying power. There is no proper epithelium on the surface, but the basement membrane (a) is very distinctly marked, while, underlying this, is the inner fibrous coat (b). Pigment never seems to get through these, and the cause of this is evidently the tough character of the basement membrane. It appears to be just as impenetrable

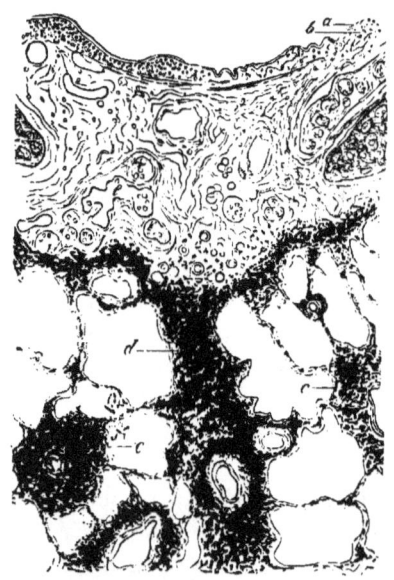

Fig. 30.—Bronchus—coal-miner's lung. × 50 diams. a, Basement membrane; b, inner fibrous coat; c, pigmented nodules surrounding arteries; d, pigmented lobular septum; e, empty mucous gland.

to bodies passing through it into the deeper parts of the mucosa as it is in allowing cellular structures to pass from the mucosa into the bronchial lumen. Were it not for the basement membrane, there is no reason why the particles should not have ready ingress to the deeper parts of the mucosa. The epithelium protecting it is never complete, and one would naturally expect that these sharp spicular bodies (Fig. 31) would fix upon the mucosa and make their

F

way into it. I have invariably found, however, that the bronchi do not become pigmented until they are so much reduced in size that no basement membrane can be recognised upon them. The greater part of the pigment does not surround bronchi, but, contrary to what one might expect, it is most abundant *round the branches of the pulmonary artery*, where it accumulates in enormous masses (*c.* Fig. 30). It seems to pass very quickly from the air-vesicles to the adventitious coat of the artery, and then becomes aggregated into more or less rounded deposits. It also accumulates in large quantity *outside* the bronchus, between the outer fibrous coat and the neighbouring air-vesicles. This is represented in Fig. 30, but, as will be noticed, it undoubtedly does not reach this situation through the mucous membrane, as the latter is quite devoid of pigment, but apparently passes into it from the walls of the air-vesicles. Another situation in which intense pigmentation is to be seen is in the lobular septa (Fig. 30), where the inhaled dust forms a deposit most intense on each side. The deep layer of the pleura is always pigmented to a high degree, but I have never seen in the superficial layer any pigment other than what may have accidentally got there in preparing the section. It is always clear and fibrous, while the deep layer is so distended with black particles, that its fibres can be with difficulty recognised. The parts which become pigmented, therefore, in spurious melanosis, are the air-vesicles, the adventitious coat of the arteries, the lobular septa, the deep layer of the pleura, the adventitious coat of the bronchus, and, ultimately, the bronchial glands.

Fig. 31.—Particles from coal-miner's lung. × 480 diams.

The course which the inhaled particles pursue is evidently the following: they first penetrate through the infundibula, air-vesicles, and smallest bronchi, and from these they are poured into certain of the peribronchial lymphatics, but mostly into the large perivascular branches. Those which enter the peribronchial lymphatics apparently very soon pass into the larger lymphatics round the arteries. From the perivascular lymphatics they run into the lobular septa, from this to the deep layer of the pleura, and, finally, into the bronchial glands. The nodules which are to be seen with the naked eye are due to pigment accumulation round small branches of the pulmonary artery, and the bluish-black colour which the surface of the pleura has, in contradistinction to the "coal-black" colour of the interior of the lung, is owing to the superficial layer of the pleura not being pigmented. This description corresponds with the course of the lymphatic vessels in the lung as described by Klein (*Lymphatic System of the Lung*), so closely, that we are bound to conclude that the pigment, after being inhaled and taken into the lung tissue, forms a complete injection of its lymphatic system. It does not follow an irregular course, but is always found in the situations above indicated. The reason why the superficial layer of the pleura is not pigmented, is that it contains the lymphatics proper of the pleura, while the deeper layer contains lymphatic vessels which are directly continuous, by means of the lobular septa, with the large perivascular branches.

If then the bronchial mucous membrane is not pigmented, if the particles which are inhaled do not pierce into it, how does it happen that chronic bronchitis should be so common among the subjects of this class of diseases? I think that the cause will be apparent on a little consideration. In the centre of each pigment nodule (Fig. 30) there is a branch of the pulmonary artery which is more or less compressed according to the amount of pigment around it, the lumen being sometimes obliterated from this cause. There must consequently be serious interference with the blood-circulation, and hence a certain amount of œdema will result. The lymphatics of the lung are at the same time choked with the foreign matters contained within them, so as also to be obliterated or seriously obstructed, and consequently the lymphatic circulation will be abolished or much impeded.

This will still more favour an œdematous condition of the interstitial fibrous tissue of the organ, especially of the loose fibrous tissue of the mucous membrane of the bronchi. As a consequence, the epithelium lining the mucous membrane will be kept in a state of constant irritation, the deep layers germinating and usually being shed long before they have completed their full development. The immediate cause of the bronchitis is undoubtedly this epithelial germination, but the remote cause is to be sought in the impediment to the blood-vascular and lymphatic circulations from pigment deposit, inducing an œdematous state of the bronchial mucous membrane, and consequently unfitting it for the growth of a proper epithelial covering. One can therefore understand why the secretion in the bronchi of catarrhal and pigmented lungs should be of a yellow colour. The colour is given to it by the abortive epithelial (or catarrhal) cells which it contains, and these, we have seen, are not directly dependent on the pigment particles for their origin.

Bronchitis associated with Chronic Disease of the Kidney.

It is a matter of common experience that those who suffer from chronic interstitial nephritis, especially where associated with gout, are prone to intercurrent attacks of bronchitis. In my examination of such cases, where there was extreme atrophy of the kidneys, I could find nothing special in the appearances, unless apparently great hypertrophy of the muscular coat of the small arteries. The appearances, otherwise, did not seem to differ from an ordinary idiopathic case.

On Chronic Interstitial Pneumonia as a Complication of Bronchitis.

Having described, at some length, the lesions which are produced in the bronchi as a result of bronchitis, I propose, next, to follow out some of those pulmonary affections which so frequently complicate a simple bronchial catarrh. I believe that one of the most important is chronic interstitial pneumonia, and as its description forms a fitting sequel to what has preceded, I shall now take it into consideration, and endeavour to show in what way it is associated with bronchitis. I believe that the relationship between bronchitis and interstitial pneumonia has not been sufficiently dwelt upon either in a clinical or in a pathological point of view. The two frequently co-exist, and I trust to show that the former in a great many cases is to be regarded as the *raison d'être* of the latter.

The disease has been known from Laennec's time by many names, all of which I believe refer to the same morbid process, and even in modern times there still seems to be very great want of clearness in the ideas of physicians as to what the disease essentially is. I find, for instance, that in Reynolds's *System of Medicine* a disease is described by Dr. Wilson Fox as "chronic pneumonia," which seems to be essentially the same as that entitled "cirrhosis" by Dr. H. Charlton Bastian in the same work. Physicians have, moreover, dwelt upon many other pulmonary affections, and given them distinctive names, which I, looking at the matter from a pathological point of view, would include under the generic name of chronic interstitial pneumonia. The disease has certain constant anatomical characters, but its causes are various; and in order to get any clear idea of what it is, we must look at it dissociated from its many points of origin. I have found that from whatever source it originates the morbid appearances and the essential process by which these are produced are invariably the same. It is apparently always a secondary affection, and mostly so to inflammatory processes in the bronchi and air-vesicles. It is undoubtedly inflammatory in its nature, all the changes met with in it being characteristic of those found in artificially inflamed parts. Its results are many and disastrous; and it is these, which by their great variety, and apparent incongruity, have to a certain extent led to the confused ideas of its essential nature.

We are more particularly concerned at present with that form of it which arises from a long-continued catarrhal bronchitis; and my description of it is drawn from a series of cases in which, undoubtedly, the commencement of the disease was a bronchial catarrh, and in which, in many instances, the existence of an interstitial pneumonia was not suspected during life.

The disease, when arising from a bronchitis, is usually bilateral, and is accompanied by diminution in the volume of the chest and marked flattening under the clavicles. There is sometimes slight dulness on percussion, but this is by no means essential. It runs a very chronic course, and may be met with in childhood, youth, or old age, but oftenest in persons of middle life. It may simulate a catarrhal phthisis in the symptoms and physical signs associated with it; but in their main features the symptoms correspond more closely with those of bronchitis. The visceral and costal layers of the pleura are invariably adherent. This union is usually fibrous and inseparable; in some cases, however, the two surfaces can be torn asunder, the bond of union being apparently of a fibrinous nature. If the visceral pleura be carefully examined, several grey and gelatinous-looking nodules, of a round shape, and each about the size of a mustard seed, will be found situated in its deep layer, provided that they are not obscured by the fibrous tissue which frequently adheres to it. On being touched they move easily with the pleura, and are with difficulty destroyed by pressure. The lung is greatly shrunken in parts, and emphysematous in others, the emphysematous parts projecting from the surrounding condensed lung-tissue. It feels hard and leather-like: it has, in fact, much the consistence of a cirrhotic liver, with, of course, the exception of the vesicular character given to it by the air which it contains. The pleura is very much thickened, and, in some places, almost cartilaginous in appearance and consistence. It may happen that calcification has occurred at certain parts, but this is not of so common occurrence in interstitial pneumonia arising from bronchitis as where it has been secondary to a pleuro-pneumonia. Considerable puckering is very frequently noticed at the apex. When the organ is incised, the thickening of the pleura can be well seen, and it can further be observed that fibrous bands, of varying thickness, and corresponding in position to the lobular septa, run inwards, and usually end by becoming attached to the wall of a cavity in the lung substance. This cavity has all the characters of one of the forms of dilated bronchi. Its shape is irregular, and it is pulled out into angular processes, so that the opposite walls may actually be in contact at certain parts. It possesses a smooth lining membrane, which has commonly some yellowish pultaceous-looking material adhering to it. One or more large bronchi communicate with it, and their mucous membrane is continuous with its lining membrane. Several fibrous bands, similar to those described above as passing inwards from the pleura, run off from this bronchi-ectatic cavity in different directions and lose themselves in the condensed lung-tissue. The lung has lost a great part of its vesicular character, and in many large areas it seems to be so closely surrounded by the thickened pleura that expansion is impossible; while, at other parts, where the thickening of the pleura may happen to be less, an apparently compensatory ectasy of the air-vesicles has occurred. Nodules of two kinds are usually to be found in the thickened bands of fibrous tissue. The one is *grey in colour*, round in shape, gelatinous in consistence, and about the size of a mustard seed: it resembles in every particular the grey gelatinous nodules referred to above as being sometimes seen in the pleura. The other is *yellow and caseous*, with frequently an ulcerated-looking cavity in its interior. It is much larger than the other—from the size of a small pea to that of a marble—and is either round or somewhat sinuous in outline. The large nodules occur irregularly throughout the fibrous bands; the small nodules run in lines outwards towards the pleura.

Some of the compressed air-vesicles contain yellowish-looking solid contents.

Following the course of the fibrous thickenings, quantities of black pigment may be seen running towards the pleura, so that in many places the fibrous tissue has a dark grey colour.

The course which the fibrous new formations, grey nodules, and pigment-particles uniformly take, is *towards the pleura.*

The peribronchial fibrous tissue, even where the bronchi are not converted into bronchiectatic cavities, is increased in quantity. The mucous membrane of the larger bronchi is much thickened, thrown into longitudinal folds, and deeply congested. In certain places the bronchial lumen seems to be of natural size, and in others it actually appears to be smaller than usual. Where this latter condition is present, the thickened mucous membrane will be observed to be the cause of it. The bronchi invariably contain yellow purulent-looking catarrhal fluid, and if some of this be examined microscopically, it will be found to contain catarrhal cells of different shapes and sizes, along with some columnar epithelium. The mucous membrane is red in colour from the congestion of its blood-vessels, and always presents the smooth glossy character given to it by the basement membrane. The bronchial glands are enlarged, indurated, darkly pigmented, and frequently caseous, the caseous degeneration being often better seen in those which are within the lung than in those outside of it.

In my description of the histological structure I shall commence with the pleura, and trace the changes in the lung substance progressively inwards. A drawing is given in Fig. 32 of such a lung, from a section made through the superficial

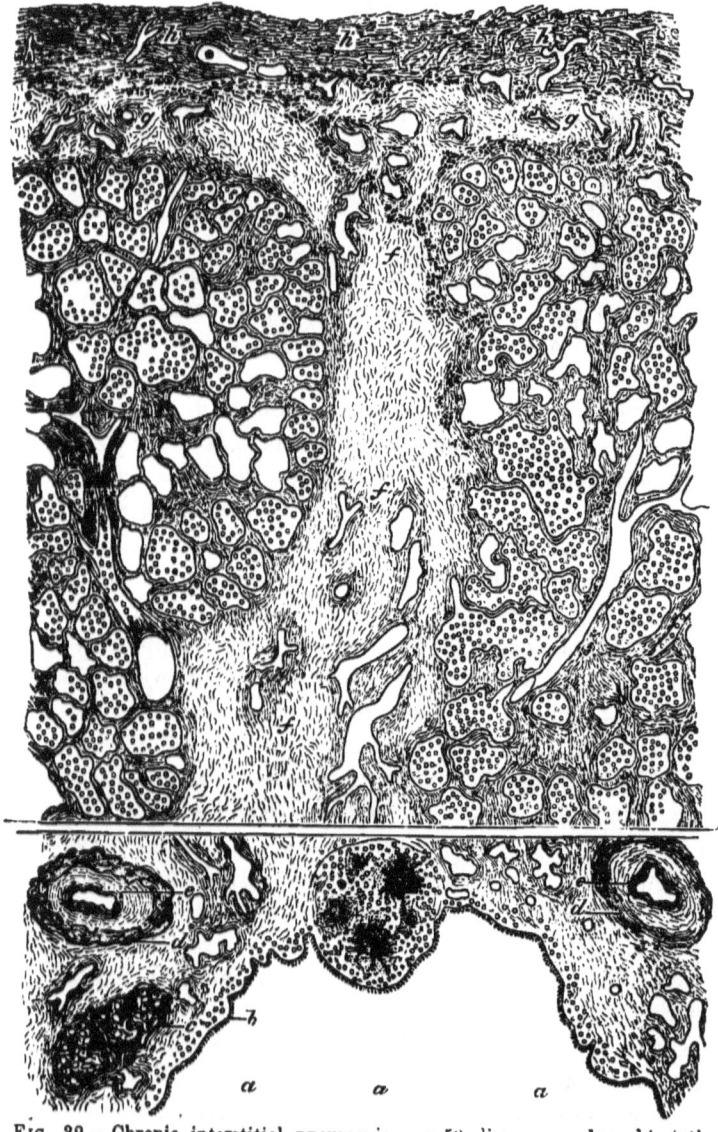

Fig. 32.—Chronic interstitial pneumonia. ×50 diams. *a*, bronchiectatic cavity; *b*, basement membrane and remains of epithelium; *c*, cartilage being absorbed; *d*, muscular coat of partially obliterated artery; *e*, intima of same; *f*, thickened lobular septum; *g*, deep layer of the pleura with pigment particles in it; *h*, superficial layer of the pleura, its adhesions torn asunder; *j*, tubercle in the bronchial wall.

layer of the pleura (*h*), down to a bronchiectatic cavity (*a*). The whole extent is not shown, the actual distance between the pleura and bronchiectatic cavity being about half as long again as is represented, the break in the figure indicating where the deficiency occurs.

At the upper part of the figure the thickened pleura is seen, torn from its attachment to the costal layer, the irregular free surface indicating the points of detachment. It still possesses two layers sufficiently well marked to be easily distinguished, and it will be noticed that both have undergone great fibrous thickening. The superficial layer (*h*) is separated from the deep layer (*g*) by a number of pigment particles, which have accumulated in the latter, but have not reached the former, an appearance which has been explained in the foregoing article on "Bronchitis due to inhalation of foreign matters." As we examine the deep layer, a thick fibrous band (*f*) can be noticed coming off from it, and passing into the lung-substance; and when traced downwards it is seen to become connected to the large bronchiectatic cavity (*a*). It is firmly united to it, and the fibrous tissue of which it is composed spreads out so as partially to surround the bronchiectatic cavity. That this large cavity is a dilated bronchus is indisputably proved by the fact of its being lined with epithelium of a columnar type (*b*), and that this rests on a thin basement membrane. The fibrous band above described is a lobular septum, and an idea may be formed of the amount of thickening which it and the pleura have undergone by comparing the appearance it presents with that of the normal lung in Fig. 1. In the two figures the connections of the pleura, lobular septum, and bronchus are alike; their relative proportions, however, are greatly altered.

On one side of this thickened lobular septum the air-vesicles are noticed, and around some of them a certain amount of fibrous thickening is to be seen. At this part of the lung, however, the fibrous thickening round the air-vesicles is by no means so evident as it was at some others, where the latter had become totally obliterated by the abnormal pressure of the former. Two of these air-vesicles are represented in Fig. 33, in process of obliteration, the vesicular spaces being seen at a, while the cicatricial fibrous tissue around them is represented at b. The alveolar walls are enormously thickened, from fibrous new formation; the outline of the air-vesicles is irregular; and the epithelium lining them has undergone a curious change, first referred to, I believe, by MM. Cornil and Ranvier (*Manuel d'Histologie Pathologique*), and evidently similar to that described by Friedländer as occurring in the pneumonia resulting from division of the recurrent nerves (Virchow's *Archiv*, vol. lxviii. p. 358). Instead of having the characters of an

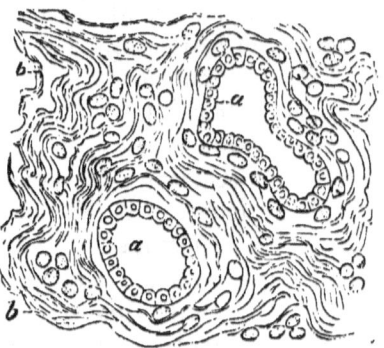

FIG. 33.—Air-vesicles in chronic interstitial pneumonia in process of obliteration. × 300 diams. a, shrunken air-vesicles lined with atypical epithelium; b, surrounding cicatricial tissue.

endothelium, that is to say a thin and delicate flat layer, it becomes cubical in shape, (Fig. 33), so as to resemble the epithelium seen on the smallest-sized bronchi; reverting, we believe, to its original type. Each cell has a distinct nucleus, and the layer, although the individual cells are much reduced in size, completely invests the alveolar cavity.

The histological structure of the thickened interlobular septa and alveolar walls varies according to the duration of the interstitial pneumonia. In its earliest condition the appearance represented in Fig. 21, gives an idea of what is to be seen. The flat connective tissue or endothelial cells lying on the bundles of fibres of the interlobular septa, and peribronchial and periarterial fibrous tissue, begin to germinate as previously described, and, as a result of this germination, large areas filled with small round cells are produced. These elongate, become spindle-shaped, and soon afterwards constitute a cicatricial tissue. That such cells actually form the fibres of the growing cicatricial tissue I am aware has been strongly controverted by many eminent authorities, but, notwithstanding all that has been said, I cannot help believing that in the *adult*, as a result of such an *inflammatory process*, they actually become transformed into the fibrous tissue. Besides these appearances, however, large masses of fibrinous lymph are commonly met with in the cicatricial tissue, apparently formed subsequently to those primary changes in the fibres. The same appearances are noticed in the periarterial and peribronchial fibrous tissue; and long rows of small round cells can be traced running from the latter into the thickened lobular septa, while, side by side with these, black particles of carbon are noticed.

Most of the parts which show cicatricial thickening have, as a rule, an abundant blood-vascular supply, forming in some places a dense cavernous-looking plexus. This, of course, is necessary for the nourishment and organisation of such highly protoplasmic structures, and is a condition by no means confined to the lung, but is met with in similar processes occurring in the liver and kidney. In many of the smaller arteries, however, there is a lesion, which I have always found associated with chronic interstitial pneumonia, and which perhaps might account for many anomalous appearances seen in them. It is that extremely important disease of the inner coat of the artery called "arteriitis obliterans" by Heubner and Friedländer, and which is referred to by the latter in the article quoted above, as ensuing in the lungs of animals in which the recurrent laryngeal nerves had been divided. It is of so frequent occurrence and generally so widely spread, that I cannot help believing that it plays a most important rôle in bronchitis, whenever there is any consecutive interstitial pneumonia present. In Fig. 32 at *e* the thickened *intima* of one of the branches of the pulmonary artery is represented as seen with a low magnifying power. The thickening is uniform, and the lumen of the vessel is much reduced in size, while the elastic laminæ do not seem much altered. In Fig. 34, the same vessel is seen more highly magnified; and it will here be noticed that the thickening is entirely due to the formation of a tissue, indistinguishable from an ordinary cicatricial production, in the *intima*. I have never noticed that it becomes atheromatous, in this respect differing from

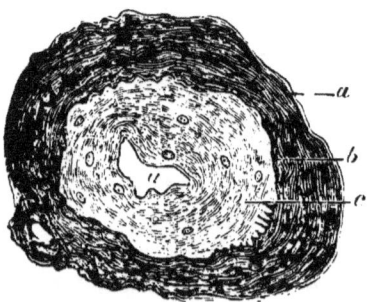

FIG. 34.—Partially obliterated artery in chronic interstitial pneumonia. *a*, contracted lumen; *b*, elastic laminæ; *c*, thickened intima.

the ordinary, and much more common, atheromatous arteriitis. The ultimate result of the process is, that the lumen becomes obliterated and the vessel converted into a fibrous cord. As a result of this obliteration, two effects are conspicuous. The one is caseation and softening of certain tracts supplied by the affected vessel, and the other is the capillary dilatation referred to above. The former of these corresponds to the large yellow and cheesy-looking deposits in the cicatricial tissue, noticed in the naked-eye description. The tracts which become caseous are either round or sinuous in shape, and the first indication of their being in process of necrosis is, that the cicatricial tissue becomes dusky in appearance, and of a brownish colour with transmitted light. A line of demarcation then forms between them and surrounding parts, and the whole affected area becomes granular and caseous. They may remain in this caseous state, or may soften in the centre and form cavities. The cavities enlarge and have ragged edges, and in this condition are frequently mistaken for bronchi undergoing ulceration. I have little doubt that most of those cases described as "ulcerative bronchitis" have got their name from the presence of cavities formed in this manner. The other result I have referred to already, as seen in the abnormally large size and number of the capillary blood-vessels ramifying in the cicatricial tissue. It is easily explained on recognised principles. Certain vessels become obliterated and the anastomotic branches enlarge to compensate for this, giving rise to an almost nævus-like dilatation in some parts. The same cause—the obliteration of the small arteries—seems to react upon the right ventricle of the heart, and induces the hypertrophy of it which is so commonly associated with chronic interstitial pneumonia.

We have seen what the cause of the large caseous nodules previously referred to is; they are portions of tissue which have undergone necrosis from the obliteration of the artery supplying them. I have, however, pointed out that there are nodules met with in this disease much smaller in size, situated either in the neighbourhood of the former, or in the pleura, and which have a grey colour and gelatinous or cartilage-like aspect. What are these? I have not the slightest hesitation in saying that they are tubercles. It would be out of place to enter into the minute description of the structure and formation of tubercle at present, but my ideas on the subject will be sufficiently understood for the purpose I have in view, if I state that a tubercle, it matters not in what organ or tissue it is situated, always has, when fully formed, the same structure. This structure is one of the best marked of all the neoplasms, and can be recognised as that of a tubercle in from two to three weeks after it has commenced to grow. It consists, to be brief, of a markedly rounded growth, quite isolated from those in its neighbourhood by means of a fibrous capsule which surrounds it, and limits its dimensions. It may be compared to a mustard seed in size. Its structure is comprised within the fibrous capsule, so that, in many respects, it is analogous to an encysted parasite. It consists essentially of one or more large giant cells, usually placed centrally, from whose periphery numbers of processes radiate outwards, which, by dividing and subdividing, constitute a reticulum, in which leucocytes with small giant cells are more or less abundantly contained. The reticulum terminates by ending in the concentric fibrous capsule at the periphery. I will not discuss the opposing views of M. Charcot and others in regard to this, but simply state that having *constantly* found the structure above described in those neoplasms known as miliary tubercle, from the same or neighbouring organs in which a *softening caseous mass* is situated, and never having found the same structure in any other body, I am constrained to conclude that there is a something having the above structure, which differs from all ordinary inflammatory products, and for which, accordingly, I retain the name of tubercle. I prefer employing this name exclusively for those structures in contradistinction to all others found in the lung, because it was originally given to them on the supposition that they were identical with those tumours found in the liver, peritoneum, and other organs, in general tuberculosis. They are identical, and I would consequently exclude all other merely catarrhal or necrotic caseous products from this nomenclature because they do not present this structure, and prefer to call them "caseating catarrhal pneumonia," or "caseous masses," according to their essential pathology.

With this preliminary statement it will be understood what I mean when I state that tubercles are found with great constancy situated in the peribronchial tissue, lobular septa, and deep layer of the pleura. In Fig. 32 one of these is represented, situated in the wall of a bronchiectatic cavity, a situation in which they are perhaps most commonly found. It is represented as seen with a low magnifying power, and many parts of its structure are, of course, consequently invisible. The giant cells, however, are clearly seen, and the general rounded isolated aspect that it possesses is also apparent. Those situated in the pleura, in interstitial pneumonia, are placed in its deep layer, while those in the lobular septa are usually met with along their margins. The caseous sources of infection in the production of these tubercles are generally numerous in chronic interstitial pneumonia, and may be either some degenerated catarrhal products in the air-vesicles or bronchi, or what is more common, some of those softened caseous masses resulting from the obliteration of the small arteries. From the situations which these tubercles occupy, manifestly around bronchi and arteries, at each side of the lobular septa, and in the deep layer of the pleura, it seems extremely probable that they originate in the lymphatics ramifying in these parts. There is much other evidence to support this view, which I cannot adduce here at present, clearly showing that the first thing noticeable in the evolution of these tubercles in interstitial pneumonia is the formation of a giant cell from lymphatic endothelium, and that this gradually throws out processes, becomes encysted, and in time attains full development.

The bronchi which have become bronchiectatic in such cases, are usually markedly deformed. When they are transversely cut, numbers of pointed cavities with sinuous margins are seen, and into the lumen there project peninsulas of the bronchial wall, some of them having a villous or club-shape. These are apparently portions of the bronchi which have been left when the retraction took place, and sometimes the retraction is so great that they appear to be attached to the bronchus only by the narrowest basis. The dilated bronchial wall is often intensely congested, and the swollen capillaries project in loops into the lumen. The basement membrane never seems to be absent however great the bronchial dilatation may be, and it is undoubtedly this which gives the bronchiectatic cavity its smooth appearance. There is always a layer of epithelial cells on its surface, usually more or less transitional in character. It is represented in Figs. 32 and 35. Sometimes the epithelium found in a bronchiectatic cavity, especially where there is much retained catarrhal secretion, is most typically columnar and ciliated, in fact, much more highly developed than one generally finds in those bronchi which are not dilated. The bronchiectatic cavities usually have some pultaceous-looking material adhering to their walls, and this, on microscopic examination, proves to be catarrhal secretion, with numbers of shed columnar cells undergoing fatty degeneration. The side of such a bronchiectatic cavity is seen in Fig. 35, and it will be noticed that

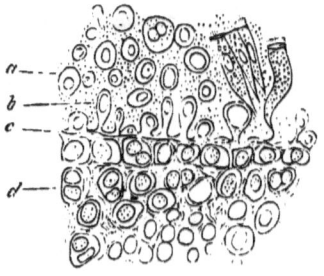

Fig. 35.—Wall of a bronchiectatic cavity in chronic interstitial pneumonia. *a*, cellular débris in cavity; *b*, columnar and transitional epithelial cells; *c*, basement membrane; *d*, cellular infiltration of wall.

there are still adherent to it some perfectly formed and ciliated columnar cells. Immediately under the basement membrane of such a bronchiectatic cavity there is usually a dense deposit of small round cells in an embryonic condition, lying in a more or less complete fibrous stroma.

The bronchial cartilages are invariably in a state of atrophy, and are undergoing absorption (Fig. 27), the process by which they are removed being that already described. Muscular fibres are not usually found in the walls of such bronchiectatic cavities. They seem to be entirely removed at an early period of the disease, apparently by atrophy from cicatricial pressure.

The air-vesicles sometimes contain a little secretion, it may be inhaled from the bronchi, but at other times probably formed within them. My reason for supposing that it is sometimes formed within them is that it presents, on certain occasions, a croupous and not a catarrhal character. The cells within it are small and round, and they are imbedded in a mass of fibrin. This, of course, is only what would be expected with such an amount of interstitial inflammation going on in the neighbourhood. Caseation is sometimes met with in the accumulated secretion, but this is comparatively unusual.

The lymphoid deposits which are found round the large bronchi, and which I cannot look upon otherwise than as small lymphatic glands, are in some cases the subject of great inflammatory infiltration, proving in certain instances too much for the nourishment supplied to them, so that caseous degeneration ensues.

I have now described the most striking and essential features of chronic interstitial pneumonia arising in connection with bronchitis, and it must next be shown what mutual relationship they possess. In order to understand this we must revert to the description given of the lymphatic system of the lung, as seen in the natural injection which takes place in anthracosis and other forms of pneumokoniosis. Foreign matters, when inhaled, pierce through the alveolar walls and inject the plasmatic spaces which, as demonstrated by Klein and others, lie upon them. From this they pierce into the peribronchial and perivascular lymphatics, distending the large lymphatic vessels of the latter to an enormous extent. They are then conducted into the lobular septa and deep layer of the pleura. The peribronchial lymphatics which lie in the adventitious coat have a free communication with the rich plexus of lymph spaces and lymphatic capillaries which are found in the bronchial mucous membrane, and the peribronchial lymphatics seem to anastomose freely with the periarterial. The course of the lymph-stream is from the mucous membrane of the bronchus to the peribronchial tissue, from this to the arterial adventitious coat, then to each side of the lobular septa outwards to the deep layer of the pleura, and hence to the bronchial glands at the root of the lung. These, curiously enough, are the situations in which the fibrous thickening is perceived in chronic interstitial pneumonia; the fibrous new formation embodying the pigment particles which are usually simultaneously present in the lung. Tubercles are formed in the same situations, and hence I am abundantly justified in saying that the interstitial thickening follows the course of the lymphatic vessels.

I have previously drawn attention to the enormous cellular deposit that occurs in the mucous membrane in chronic bronchitis, and how it seems to be retained within the bronchial wall by means of the strong elastic basement membrane lining the mucous surface. We have, moreover, seen that, from the bronchial mucosa, long lines of the inflammatory cells infiltrating it stretch into the bronchial adventitia and lobular septa. They follow the course of the lymph-stream, and undoubtedly, like the inhaled particles which accompany them, find their way to the deep layer of the pleura, and to the bronchial glands. From this, the cause of the interstitial pneumonia in chronic bronchitis will be apparent. The accumulation of inflammatory effusion which takes place in the mucosa is pent up by the resistent basement membrane. It cannot be thrown off into the bronchus, and, consequently, in the course of time, the cellular structures, partly by their amoeboid movement, and partly by the force of the lymph-stream, pierce into the peribronchial and periarterial lymphatics, and not only infiltrate but also excite them to inflammatory action, resulting in the production of the changes seen in Fig. 21. The same thing happens in the lobular septa. The lymphatics become distended with inflammatory cells, irritation is the result, and proliferation of endothelia and blood-vascular diapedesis follow. The pleurisy which ensues in such cases is undoubtedly due to the same cause. The deep layer of the pleura in an early stage of the disease can be seen to be enormously infiltrated with inflammatory exudation, there is great vascular disturbance, and, apparently, the superficial layer participates in this, and exudation of plastic lymph on its surface results. All the subsequent changes in the pulmonary interstitial tissue and pleura can be easily understood when the cause of their origin is borne in mind.

Looking at all this in a comparative point of view, there is nothing very extraordinary or unusual in it after all. It is nothing more than what is seen in a limb, for instance, where an irritation of a peripheral part has existed for some time. In such a case the lymphatics of the limb become in course of time enormously thickened, and a permanent development of cicatricial tissue around them results. In the lung, in bronchitis, we have analogous conditions. The catarrhal inflammation is a constant source of irritation in the mucosa, and, as a consequence, the lymphatics in communication with it, having to discharge the part of removing unnecessary tissue products, become clogged and themselves irritated. Their irritation reacts upon neighbouring parts, and induces the inflammatory changes to which I have just referred.

It is by no means necessary that the irritation should always primarily be bronchial. A degenerating deposit in the air-vesicles will, in course of time, produce the same lymphatic disturbance in varying degree, according to the amount of irritation present. All caseating masses of catarrhal products accumulating in the air-vesicles cause a certain amount of chronic interstitial pneumonia, probably from the absorption of the caseous material; and whether the interstitial pneumonia results from such accumulations in the air-vesicles, or whether it arises in consequence of bronchitis, the effects are *essentially* the same, although the thickening in the one case may primarily be localized to a certain group of air-vesicles, and in the other to the bronchus in which the catarrh has originated. Both ultimately tend to thickening of the lobular septa, to a chronic form of pleurisy, and to an enlargement of the bronchial glands—a course that is fully explained by the ramifications of the pulmonary lymphatics.

Ever since Laennec showed that many of the cavities found in the lung in various diseases were in reality dilated bronchi, numerous theories have been offered in explanation of how they are produced. It seems to me that in most cases the inventors of these theories have been much too exclusive in their reasoning, and have too often ridden their pet "hobby" to death in endeavouring to explain effects which obviously may originate from various causes. I do not intend to consider the subject of bronchiectasy in a systematic manner at present, but will merely refer to the special form of it which is connected with chronic interstitial pneumonia. I have described the appearances in these bronchiectatic cavities, and it remains to be shown in what way the dilatation is brought about.

Corrigan, in his classical essay on the subject, says:[1] "If there were but one bronchial tube with contracting fibrocellular tissue around it, then the contracting tissue would, as in the instance of the stricture of the œsophagus or rectum, cause narrowing of the tube; but when there is, as in the lung, a number of bronchial tubes, and the contracting tissue not placed around the tubes but occupying the intervals between the tubes, then the slow contraction of this tissue will tend to draw the parietes of one tube toward the parietes of the other, and necessarily dilate them." To this statement, with all due deference to its eminent author, I am constrained to say, "*non sequitur*." For if it be true, why then is it that in cirrhosis of the liver, and in cirrhosis of most of the other organs, dilatation of the tubular structures and blood-vessels within them does not occur from the same cause? There is a slip in the author's statement which probably has been recognised since the above was written, and which consists in the omission of the fact that the rigid chest-wall acts as the fixed point in the dilating process. The liver, œsophagus, rectum, and most other organs, have no

[1] *Dublin Medical Journal*, 1838.

such fixed surroundings to form a point of resistance, so that in cirrhosis of them, contraction takes place towards their own centres, and produces constriction and narrowing of all the tubular structures within them. The two pleural surfaces of the lung, however, may, practically speaking, be said to be united, and the invariable pleurisy which accompanies chronic interstitial pneumonia of course renders this complete, so that there is the rigid arch formed by the ribs and vertebræ to act as a counterpoise to the easily dilatable bronchial-wall. I shall presently show that the dilating power is in great part to be found in the contracting fibrous tissue, and will be exerted in a direction towards the central point of the contracting band, reacting on the one hand on the chest-wall, on the other on the bronchial-wall. A balance is then to be struck in the effect which this traction will produce when exerted equally upon the arch of the comparatively thin bronchial-wall in a direction outwards, and on the rigid costal-wall in a direction inwards. The result that must follow is obvious. Both the chest-wall and the bronchus will be influenced, but the former, being much the stronger of the two, and representing an arch with its *concavity* towards the point of traction, will be influenced to a less extent than the thin wall of the bronchus, which, moreover, represents an arch with its *convexity* towards the point of traction. The bronchus consequently becomes immensely dilated and the chest-wall to a certain extent retracted—two phenomena which are of constant occurrence in this disease. In fact, the bronchial dilatation would be very much greater than it usually is were the loss of volume in the chest contents, through cirrhotic contraction of the lung, not to a certain extent compensated by the drawing upwards of the diaphragm and liver.

But let us examine a little more closely the special mechanism which exists in the lung for the carrying out of these mechanical data. If we look at Fig. 1 it will be seen that, continuous with the deep layer of the pleura, the lobular septa run into the lung-tissue, and are directly united to the wall of the bronchi. There are usually, as represented in the figure, three or more such lobular septa attached at intervals to different parts of the bronchial circumference, and they are inseparably united to the peribronchial fibrous tissue. When these lobular septa are traced throughout the lung, in whatever direction they pass off from the bronchus, they invariably end, it may be after giving and receiving many collateral branches, by becoming united to the deep layer of the pleura. There exists accordingly, in the normal

FIG. 36.—Chronic interstitial pneumonia. × 2 diams. *a*, thickened pleura; *b*, bronchiectatic cavity; *c*, thickened lobular septum; *d*, thickened septum running inwards.

lung, a mechanism which is specially adapted for the production of bronchiectatic cavities on the principles I have referred to.

Let us see whether there is any direct evidence that the shape of the bronchiectatic cavities corresponds to the lines in which the dilating force has been applied. Fig. 36 represents a large section of a lung in chronic interstitial pneumonia magnified about two diameters, the same lung in fact as that represented more highly magnified in Fig. 32. The cavity in the centre is a bronchiectatic cavity, part of which is also represented in Fig. 32, and around it are noticed the thickened bands of cicatricial tissue, while the pleura, also much increased in bulk, is seen at the upper part of the figure. The cavity, it will be noticed, is very irregular in shape, being drawn out into angular projections. If these angular projections be looked at, it will be observed that bands of cicatricial tissue run off into the lung substance in

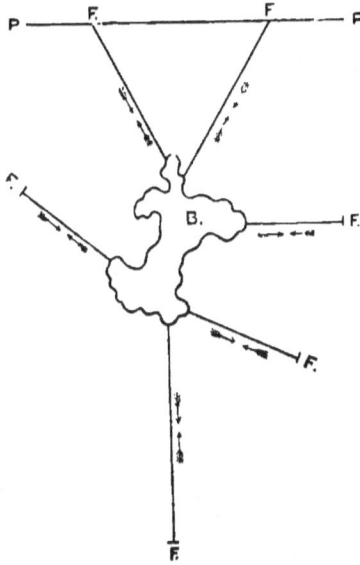

FIG. 37.—Scheme of Fig. 36. P P, pleura; F F, attached points in pleura and surrounding parts; B, bronchiectatic cavity.

lines corresponding with each. If we represent the contracting bands diagrammatically, in the scheme seen in Fig. 37, the direction of the contraction will be better understood. Let P P represent the thickened pleura, B the bronchiectatic cavity, and F F the fixed points in the pleura on the one side, and at corresponding parts of the pleura on the other. If, now, the contracting bands of cicatricial tissue shorten towards their central points, the traction will be exerted in the direction of the arrows, and the conclusion is irresistible that the bronchus, being the weaker of the two attached points, must become dilated. The irregular shape which the dilated bronchi have in this variety of bronchiectasy, as distinguished from certain other forms whose cause is different, entirely bears out what I have said in regard to them. That the (negative!) inspiratory pressure aids, as suggested by Corrigan and others, in their production, is, I think, somewhat doubtful. I would look upon the bronchiectasy, in this disease, not as one produced by sudden alterations of the atmospheric balance, but rather as illustrating a principle that is to be found in many other instances in connection with contractile new formations in different parts of the body.

On Vesicular Emphysema and Collapse of the Lung as Complications of Bronchitis.

We may safely say that if a person has suffered from chronic bronchitis, there will, as a rule, be more or less vesicular emphysema. The chest in such a case is barrel-shaped, the lungs are found to cover the heart, they have a greyish-pink colour, and the pigment which exists in all adult lungs is seen specially well between the lobules, on account of their distension and pale grey colour. The over-distended part does not crepitate so much as the natural lung, and, when cut into, is found to be anæmic, while the centre of the lung is usually deeply congested. There is not necessarily, in such a chronic case, any collapse of neighbouring portions of lung-tissue. The whole organ is usually increased in volume, but the emphysema is best marked at the apices and anterior margins.

Vesicular emphysema, however, also occurs as a complication of acute bronchitis, more frequently apparently in children than in adults, and, as Dr. Gairdner pointed out

(*Edin. Monthly Journ.* 1851), collapse of neighbouring portions of lung-tissue usually accompanies this form.

The etiology of emphysema has been matter of dispute for many years, and even yet authorities are by no means agreed as to its dynamics. Mendelssohn and Jenner seem to have been the first to suggest the "forced-expiration" theory of its production, although the possibility of an expiratory effort, as in coughing, giving rise to over-distension of the air-vesicles, is by no means universally admitted. The cough of a chronic bronchitic is a constant feature of the disease, and *à priori* seems to be a very rational explanation of the accompanying emphysema. In coughing, a full inspiration is taken, the glottis is closed, and then the muscles of forced expiration contracting upon the lung, compress the air within it and raise its tension. Dr. Gairdner, however (*loc. cit.*), compares the chest in inspiration to a bladder filled with air, and, very properly, in view of this comparison, says that if it be equally compressed on all sides no distension can occur, but, on the contrary, it will become diminished in size. This would be true were the likeness between the chest filled with air and a bladder, under the same circumstances, literally correct. When, however, we come to compare the two it is evident that the conditions are not really collateral. In the case of a hollow viscus like the bladder distended with air, and without any aperture of exit, the contents are constant, but in the case of the chest the lungs are not the only occupants, but the heart and large blood-vessels take up a very considerable space. The size of these varies, according to the amount of blood contained in them, so that a considerable space might be left in the chest which will naturally be filled by the over-expanded lung. In the circumstances that I have presupposed, in forcible expiration, the lung is distended to begin with, that is to say, it contains as much air as the full inspiratory capacity of the chest will admit. It is then forcibly compressed, and the tension of the contained air is raised, and if that were all that really happened I cannot see how the lung could possibly become over-distended, and, so far, would coincide in Dr. Gairdner's view. But is this really all that occurs? I think there is another effect of the forcible-expiratory movement which is overlooked in this comparison, namely, that the blood which is contained in the heart and large blood-vessels is at the same time driven outwards, as evinced in the distension of the temporal arteries and the venous turgescence which chronic bronchitics exhibit when coughing. This would leave a certain amount of space, were it not that the increased tension of the air within the lung drives the latter into the part where the void is left. The parts at which I would expect the emphysema to be most marked would consequently be near the heart and large blood-vessels, and as a matter of experience this is generally the case, the anterior margin being notoriously the situation in which vesicular emphysema is most often met with. The cause of its localisation to this part is generally said to be that the lung is weaker here than elsewhere, and that consequently it tends to expand more easily. I cannot see that this should be so, for the lobular septa, which practically are the means by which the air-vesicles are retained in their relative situations, are better developed at the anterior margins than elsewhere. The compressed air acts equally in all directions, and there is no reason, so far as I can see, why the anterior margins and apices should become over-distended more than any other part of the periphery, on the supposition that the cause is an expiratory effort. It seems to me much more probable that the tension of the air within the lung is first raised, then a considerable quantity of blood leaves the chest, and that part of the lung from which the support has been removed becomes over-distended. There has been taken off from this side of the lung a certain amount of the surrounding support afforded by the contents of the heart and vessels, and the air within the compressed lung being at a high pressure drives this portion into the vacuity.

There is little doubt, however, that although, in chronic bronchitis, the alteration in the bulk of the contents of the chest in forced expiration must play a most important part in originating the emphysema, yet the enlargement of the whole chest, and more especially of the intercostal and supra-clavicular spaces seen in those suffering from it, must materially aid in the process. The intercostal spaces, apparently from the inability of the intercostal muscles to withstand the pressure from within, may be seen to become dilated in an emphysematous person with thin chest-walls in each forced expiratory effort.

Collapse of the lung is also a common accompaniment of bronchitis, more especially of an *acute* attack, where the small bronchi are filled with mucous secretion, and where there is, at the same time, a certain amount of acute catarrhal pneumonia. The appearance of the lung in such a case is very characteristic. There are patches over its surface which are sunken below the level of the surrounding parts, and which have a bluish-purple colour. The parts around them are of a light pink hue, and are seen to be over-vesicular. The lung on being cut into is found to be much congested, and, when exposed to the atmosphere for a few minutes, becomes of a bright scarlet colour, contrasting in a marked degree with the dark red colour which the blood presents in *chronic* bronchitis with emphysema. The mucous membrane of the middle-sized bronchi is red, and contains a little muco-purulent secretion, but large numbers of the smaller-sized bronchi are completely occluded, from the muco-purulent discharge which they contain. It can be squeezed out from them, and is tough and tenacious, although as yet fluid. Catarrhal pneumonic patches of a greyish colour, and with an indistinct margin, are seen over the lung, and yield some of the same muco-purulent secretion on being squeezed. These, however, are not hard, the hand on being passed over them detecting their boundaries with difficulty, evidently from the as yet fluid state of the pneumonic secretion. The collapsed portions of lung in connection with these are visible on section, placed at the periphery, and involving a small bronchus and attached lobule or lobules.

The cause of the collapse in such a case is that a terminal bronchus, opening into a group of air-vesicles, becomes plugged with a mass of muco-purulent discharge. This, I have found, may be formed locally, or, as sometimes happens, may be inspired from a bronchus of larger calibre, the occluded bronchus sometimes retaining a complete layer

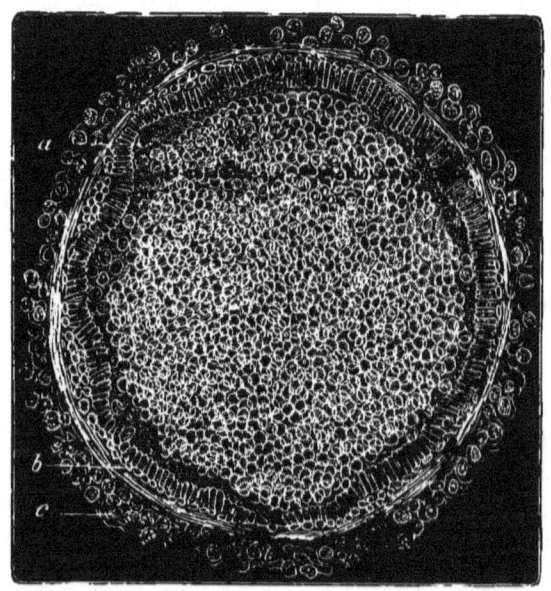

Fig. 38.—Small bronchus in acute bronchitis, occluded by a plug of inhaled catarrhal secretion. × 350 diams. *a*, catarrhal plug; *b*, epithelium lining bronchus; *c*, surrounding adventitious coat infiltrated with cells.

of columnar epithelium, and evidently not being that from which the secretion has been formed. The accompanying figure (Fig. 38) shows the occlusion which may occur. It was taken from a case of acute bronchitis, with complete collapse of the portion of lung beyond, and vesicular emphysema of the surrounding parts. It will be noticed that the plug of catarrhal secretion has apparently not been formed at this part, but must have been inspired, as the epithelium on the bronchus is still complete. The manner in which the collapse is brought about in such a case is evident. The mucus accumulates within the bronchus, it is moved to and fro by the inspiratory and expiratory efforts, but gradually becomes drawn outwards in the inspiratory act towards the periphery, and is impacted in a small terminal bronchus. The plug will allow the air to escape from the occluded portion of lung, because the calibre of the proximal end of the bronchus is greater than the distal, but, when drawn backwards in inspiration, it becomes impacted in a small terminal twig, and effectually prevents the further ingress of air. It therefore follows that the collapsed portions of lung will be at the periphery where the bronchi are smallest. The catarrhal plug acts like a ball-valve in allowing air to pass in one direction, but preventing it from passing in the other.

When the collapse is complete in several parts of the lung, then the remaining vesicular portions of lung-tissue are insufficient in bulk to fill the chest in full inspiration, and consequently become over-distended from atmospheric pressure. It is only under such conditions that the otherwise negative inspiratory force can have any action in dilating the air-vesicles. So long as the lung is sufficient in bulk to fill the cavity of the chest in full inspiration, then the inspiratory act can have no influence in causing over-distension. If the air within the lung becomes unduly stretched by the inspiratory act and its tension lessened, then the atmospheric influence comes into play, but under ordinary circumstances, the tension of the air within and without the chest being virtually equal, no disturbance of the atmospheric balance occurs.

On Bronchiectasy as a Complication of Bronchitis.

I have already described the bronchiectatic cavities which are so commonly found in chronic interstitial pneumonia resulting from bronchitis, and have shown that the cause of the ectasy in that disease is to be sought in the traction of the thickened interlobular septa upon the bronchial wall. The bronchial wall becomes distorted, pulled out into angular cavities, and greatly thickened, on account of the cellular and fibrous formation in and around it. Bronchiectasy, however, frequently occurs in connection with bronchitis, quite independently of there being any interstitial pneumonia, and, in this case, the dilated bronchi have quite a different appearance.

It commonly happens that the subject of this latter form of bronchiectasy has been a chronic bronchitic, and there sometimes is very considerable emphysema concomitantly present. The bronchial dilatations are uniform in shape, and the walls, instead of being thick and fibrous, are thin and stretched. Their shape is either fusiform or saccular, more often the former. When saccular, the dilatation seems to occupy the end of a bronchus, while the lung beyond will usually be found to be non-vesicular, from collapse or from pneumonic infiltration of the air-vesicles. Collapse along with the sacculated form of bronchiectasy is as marked a complication as emphysema is in others. When accompanied by emphysema, the dilatation of the bronchi is usually fusiform or "finger-glove" in character, the bronchus being nearly as large at the periphery as at the centre of the organ, and there is no marked obstruction at any part. I have had the opportunity of studying the saccular form in one very remarkable instance, among others, in the lung of a child who had suffered from bronchitis and acute catarrhal pneumonia. The lung on section presented all throughout its substance large round saccular bronchiectatic cavities from a quarter to half an inch in diameter, lined with a smooth membrane, and nearly empty. The air-vesicles were completely collapsed all round these, so that the lung looked more like that of a tortoise than that of a human subject. On careful microscopic examination no chronic interstitial pneumonia was found which might account for the formation of these cavities, and the uniformly round shape which they presented precluded the idea of their being thus formed. The bronchiectatic cavities, on microscopic examination, presented the usual appearances of bronchitis, but in the smaller bronchi and certain of the air-vesicles the catarrhal secretion had accumulated and produced occlusion of them much in the same way as seen in Fig. 38. I have frequently seen this sacculated form of bronchiectasy in other cases of bronchitis with collapse, but never to the same extent as in this. In the adult they usually contain intensely putrid catarrhal products, which seem to have accumulated within them after the dilatation took place. It is the rarest occurrence to find an abraded surface on one of these bronchiectatic cavities, and I must again express my belief that many of those described as ulcerative bronchiectatic cavities have in reality arisen from necrosis of solidified lung-tissue. The cause of their smooth appearance is apparent on microscopic section; it is due to the elastic basement membrane that I have so often referred to when treating of bronchitis. It is this which prevents the bronchiectatic cavity having a granulating surface, for by its elastic reaction it prevents the superficial vessels from being thrown out on the bronchiectatic wall, so as to constitute granulation loops. Were it not for this basement membrane the bronchus would be a specially favourable site for the production of a granulating surface, seeing that there is here an abundant superficial plexus of vessels, which, if not restrained, would, on the principles I have stated in my paper on the "Process of Healing,"[1] be thrown out on the surface as granulation vessels. I have never seen a true granulating surface on a bronchus in bronchitis, although I have occasionally observed the base-

[1] *Journal of Anatomy*, July, 1879.

ment membrane thrown out into villus-like processes, which may be regarded as the nearest approach to a granulating surface that is ever reached where the basement membrane is entire. The basement membrane has the same restraining influence here that the skin has on the subcutaneous vessels. Remove the skin and expose the surface and it will granulate, the reason being that the restraining action of the cuticle upon the underlying vessels has been taken away, and the blood-pressure within the latter throws them out on the surface. The same would apply with special force to the bronchi, for if the basement membrane were removed from their surface then there would be no medium of restraint to re-act upon the underlying vessels, and they would undoubtedly form granulations. The basement membrane is, however, always present, and this accounts for bronchiectatic cavities presenting a smooth interior and not granulating.

The formation of these saccular and fusiform bronchiectatic cavities with thin walls, and unaccompanied by chronic interstitial pneumonia, is matter of interesting speculation. Professor Stewart[1] supposes that there is primarily an atrophy of the wall arising from some unknown cause, and that the bronchus being thus weakened is more readily acted upon by alterations of atmospheric pressure. The accumulations of catarrhal products which take place within them he regards as of secondary occurrence, and he would look upon their presence as merely due to the retentive character of the bronchiectatic cavity. That an atrophy of the wall, and more especially of the muscular coat, is a strong predisposing cause of bronchiectasy, is, I think, indisputable. I have already shown how this may arise from the presence of bronchitis. The cellular accumulation which is met with in the bronchial wall, as a result of bronchitis, comes to exert a deleterious pressure upon the muscular fibres, and gives rise to atrophy of them. It can further be easily understood that if the muscular coat is surrounded and

[1] *Edinburgh Medical Journal*, 1867.

penetrated by this cellular effusion its function in resisting bronchial dilatation from forced expiratory efforts will be seriously interrupted.

It cannot, therefore, I think, be denied that there would be the liability in coughing to dilatation occurring in the bronchi, just as it takes place in the surrounding air-vesicles. The dilatation of the bronchus in such a case is uniform, and is accompanied by similar dilatation in the air-vesicles, and it is only natural to conclude that the one is a mere extension of the other. The agency by which the dilatation is effected is undoubtedly the compression of the air contained within the lung in forced expiration.

In accounting for the sacculated dilatations at the end of a bronchus, where there is accompanying collapse of the air-vesicles, the same explanation will not hold good, for here we have not the universal dilatation of all the air-passages, but collapse of certain parts of them, namely, of the air-vesicles. In such a case the terminal bronchi will be found to be plugged, and collapse, as a consequence, takes place in the parts of the lung into which the occluded bronchi open. The bronchial walls are, at the same time, weakened from cellular effusion into them, as bronchitis usually occurs along with the bronchiectasy. It consequently seems only reasonable to expect that when a full inspiration is taken, some of the bronchi must become dilated. The lung must expand in order to follow the enlargement in the different parts of the chest, the air-vesicles in the collapsed portions cannot expand, and the next less resistant of the respiratory passages, namely, the small bronchi, become dilated, by the atmospheric pressure.

It also seems likely that, in some instances, the mere accumulation of bronchial secretion within a bronchus may, when it expands from fatty degeneration, cause dilatation of a bronchus. This is always partial, and is usually accompanied by more or less pneumonia, and frequently by partial gangrene of the lung.

In summing up the different factors which may be

instrumental in producing bronchial dilatation I find them to be the following :—

1. The traction of cicatricial tissue on the walls of the bronchi.
2. Forced expiratory efforts.
3. Atmospheric pressure, when there is extensive collapse elsewhere.
4. Accumulation of catarrhal products within a terminal bronchus.

On Catarrhal Pneumonia as a Complication of Bronchitis.

The frequency with which catarrhal pneumonic infiltration of groups of air-vesicles follows an acute bronchitis supports the theory that a considerable amount of the pneumonic infiltration is simply bronchitic secretion which has been inspired. An identical appearance is noticed when an aneurism ruptures into the trachea and some of the blood passes into the lung. Groups of air-vesicles are seen filled with blood, and they have a shape and distribution similar to those seen in catarrhal pneumonia, the blood being substituted for catarrhal products. In this case the blood has evidently been sucked into the lung during an inspiratory effort, and there seems very good reason to believe that in acute bronchitis some of the pneumonic infiltration is inhaled in the same manner. Microscopic examination of the pneumonic infiltration favours this view, as the cellular products in many cases show that they have been derived from the bronchi.

PART II.

ON CATARRHAL PNEUMONIA AND TUBERCLE IN THE HUMAN LUNG.

INTRODUCTORY.

IT has been shown that the chief source of the cells in the expectoration of bronchitics is the deepest of the three epithelial strata which line the bronchi; while the superabundant mucous secretion is due to an increased activity of the glands of the mucous membrane, probably caused by an inordinately abundant blood-supply. The production of each constituent of the increased secretion, it was shown, consisted simply in an exaggeration of processes which occur under natural conditions. Normally, the fully-developed bronchial epithelium found in the superficial stratum originates from the embryonic layer lying below. New cells are constantly being produced in this embryonic layer by a process of division, which springing upwards, are gradually evolved into ciliated columnar epithelium. In bronchitis this division of cells in the embryonal layer goes on much faster. They are over-stimulated, and instead of the resulting new-formed elements developing into columnar epithelium, they are cast off, in an incomplete state, as the cells of the catarrhal secretion.

My object will now be in the first place to trace out, in a

similar manner, the so-called catarrhal changes which affect the alveolar walls, with special reference to their essential nature and mode of origin; and to show what the course, usual terminations, and complications of such catarrhal pneumonic processes are. A brief contrast will also be drawn between this and what is known as "croupous pneumonia"; and as the subject of "tubercle" is so inextricably bound up with that of catarrhal pneumonia in its different stages, it will also be shown what the mutual relations between it and catarrhal pneumonia are.

It is, however, absolutely necessary for the appreciation of what will follow, that the natural structure of the walls of the air-vesicles be made perfectly clear, for, we shall see that, in catarrhal pneumonia as in bronchitis, the disease is in reality one of degree, and that it is impossible to draw a hard and fast line between the condition of parts in the normal alveolar wall, and that which can be perceived when it is in a state of catarrh.

The Structure of the Wall of the Natural Air-Vesicles.

After the bronchioles have reached their minimum size, they expand into what is known as the infundibulum,—a mere common channel opening into the surrounding air-vesicles. It might be compared to a corridor with which many chambers freely communicate. It has no special walls, as in a bronchus, but is bounded on all sides by the adjacent air-vesicles.

As the bronchi diminish in size, the three strata of epithelium covering their mucous membrane are succeeded by a single layer, composed of somewhat cubical-shaped cells, and, as the air-vesicles are reached, the epithelial covering assumes all the characters of an endothelium, and by many is looked upon as such rather than as an epithelium. Such a classification, however, is misleading, for, as the epithelium of the whole of the respiratory passages has a common origin

THE AIR-VESICLES.

in the embryo, we shall find that it is important to keep this clearly in view in investigating the diseases to which it is liable.

In order to see the epithelial lining of the alveolar wall it is necessary to stain it with silver. The outline of the epithelial cell is so delicate, and so completely transparent, that it cannot be otherwise observed. The silver markings,

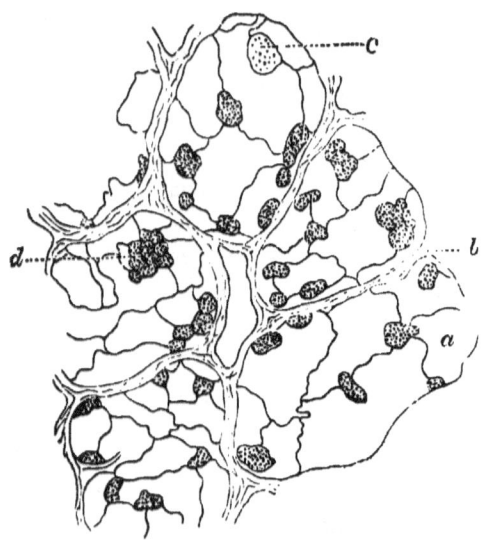

Fig. 39.—Alveolar cavities from the kitten, stained with nitrate of silver. ×450 diams. *a*, outlines of fully developed epithelial cells; *b*, alveolar walls; *c*, a young epithelial cell losing its granular appearance; *d*, a group of young epithelial cells germinating.

however, show it with the greatest precision. Two representations of it are given in Figs. 39 and 40. The majority of the cells are large flat plates with sinuous borders (Fig. 39, *a*), usually having a nucleus in the centre, which is only rendered visible, after being silvered, by the application of different colouring reagents, such as hæmatoxylon. In many instances, evidently in the oldest cells, a nucleus cannot be perceived, and such cells seem to consist merely of a flat, structureless plate of a keratine-like substance. They wind round the partitions between one air-vesicle and another, and mould themselves to all the inequalities of the surface, in this way forming a complete investment for the underlying fibrous tissue and lymphatic spaces of the alveolar wall.

Besides these, however, there are other bodies of a somewhat different structure, which are constantly seen in lungs silvered by injection into the trachea. They are more abundant in young animals than in old, and, apparently, are

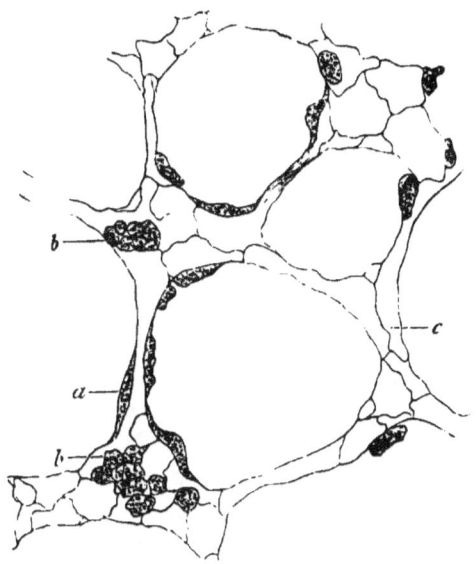

FIG. 40.—Alveolar cavities from the kitten stained with nitrate of silver. × 450 diams : giving a profile view of the epithelium.

more numerous in the lungs of some species than others. In the lung of the kitten they are particularly well developed, but in the lungs of all young mammals, including that of the child, they can be seen. They are more or less polygonal cells, which sometimes lie in little groups (Fig. 39, *d*), at other times have an isolated position (Fig. 40, *a*). When stained with nitrate of silver they have not the same homogeneous aspect as the larger cells, but appear to possess a

THE AIR-VESICLES.

much more granular consistence. They can, moreover, on account of this granularity, be distinctly seen lying on the alveolar wall, in preparations unstained with silver and simply in their natural state, or when coloured by logwood or carmine. They stain deeply with these colouring reagents, while the larger flat cells remain unaffected. When viewed in profile as in Fig. 40, *a*, they are seen not to be flat, but to rise above the surface, and to project into the alveolar cavity. They can be detached by means of pencilling, or merely by the pressure of the cover-glass, and they leave a little cup-shaped space when so removed. They sometimes exhibit a nucleus, especially if they are large, but, in other cases, this is imperceptible.

Various theories have been offered as to what these bodies are. It has been suggested by Klein (*Lymphatic System of the Lung*) that they correspond to the pseudo-stomata found in serous membranes,—the projecting cell being in reality an underlying connective-tissue corpuscle, which, pushing its way through an aperture in the endothelial lining, comes to project on the surface in the manner described. This, however, is not borne out in fact, for when we come to examine an alveolar wall which has been transversely cut, and where these can be seen in profile as in Figs. 39 and 40, it is evident that they do not sink into, or originate from, the substance of the alveolar wall, but are confined to the surface, at the same level as the ordinary flat epithelium. They, further, do not resemble pseudo-stomatous cells, in the fact that groups of them are found scattered here and there over the wall of the air-vesicle, instead of being single, as in the case of pseudo-stomata in other parts. They look much more like the young endothelial cells found on serous membranes, as, for instance, on the omentum and mesentery, and it seems only rational to conclude that they are such, and that they, in course of time, would supply the place of those older cells which are constantly being cast off. The conversion of the granular protoplasm of the one into the hyaline substance of the other can actually be demonstrated in preparations such as that from which Fig. 39 was taken, where cells, intermediate in size and granularity (c), can always be discovered.

That these embryonic epithelial cells are constantly germinating there cannot be the slightest doubt, for groups in which the cleavage is proceeding are of common occurrence, and it is a significant fact, as bearing on what follows, in regard to the predisposition of young individuals to catarrhal pneumonia, that these germinating groups are undoubtedly more numerous in young than in old animals.

It is therefore apparent that, in structure, the epithelium lining the air-vesicles has the closest resemblance to the deepest stratum covering the bronchus. Both, when silvered, have much the character of an endothelium, and both contain young germinating cells.

Although there seems to be little evidence that these young granular embryonic cells are of a connective-tissue type, yet that they have a pseudo-stomatous action in transferring foreign bodies from the alveolar cavity to an underlying lymphatic as claimed by Klein and others, can be verified. In order to understand how this takes place we must examine the other constituents of the alveolar wall. The basis or stroma of the alveolar wall in the human subject is mainly composed of yellow elastic fibres and capillary blood-vessels. The former are in abundance, and have a figure-of-eight arrangement between one alveolar cavity and another. Besides these two constituents, however, there is a little white fibrous tissue with nuclei lying upon it, and, if carefully examined, free wandering cells of large size, and probably of a connective-tissue type, are also to be seen. These cells are shown in Fig. 41, c, taken from the lung of the kitten, but they are also found in the human lung quite as abundantly. The blood-vessels of the lung from which this drawing was made had been injected with silver, and sufficient had exuded from them to stain some of the parts in their immediate neighbourhood, manifestly of the connective-tissue cells above referred to. They are large rounded, or oval bodies, which lie at a deeper level than the epithelium, and usually close to the concavity of a capillary blood-vessel, whose shape they assume. They are finely granular, and differ in contour from the germinating epithelium before described, part of which is also seen at the lowest portion of the figure. The capillary blood-vessels ramifying on the alveolar walls are represented in this figure, showing how great a part of the space between two adjacent alveolar surfaces is occupied by them. The sinuous lines seen upon them are, of course, the borders of their endothelial cells, stained with silver.

Besides blood-vessels, however, there is an abundant lymphatic supply, in the shape of plasmatic spaces, which lie between the bundles of elastic and white fibrous tissue. They form a series of intercommunicating and branching channels in the substance of the alveolar wall. In order to see them they must either be stained with silver or, what is much better, examined in the lung of a coal-miner. In the latter case they become injected with inhaled particles of carbon or coal, which indicate their shape and position with great distinctness. A representation is given in Fig 42 of a transverse section of an alveolar wall, from a well-marked case of anthracosis. On its surface a group of germinating epithelial cells, (*a*), similar to those seen in Fig, 39, *d*, is represented, and other groups are seen at different parts of the figure, in neighbouring air-vesicles. They are raised above the surface, and, in the interspace between two of them, black pigment particles have insinuated themselves, and have been carried into an underlying plasmatic or lymphatic space (*b*). Here they have accumulated, and have injected the space, so as to show its contour. These spaces have a branching or stellate appearance, and it can be seen how the pigment contained in the one is transferred to the other, by means of the free anastomosis which exists between them. In this way a complete series of lymphatic channels is formed in the alveolar wall, having a direct communication with the alveolar cavities by means of the openings between the epithelial cells. Foreign bodies are carried by them into larger lymphatic vessels in the bronchial and arterial adventitious coats, and also into the interlobular septa, by which they ultimately reach the bronchial glands and deep layer of the pleura.

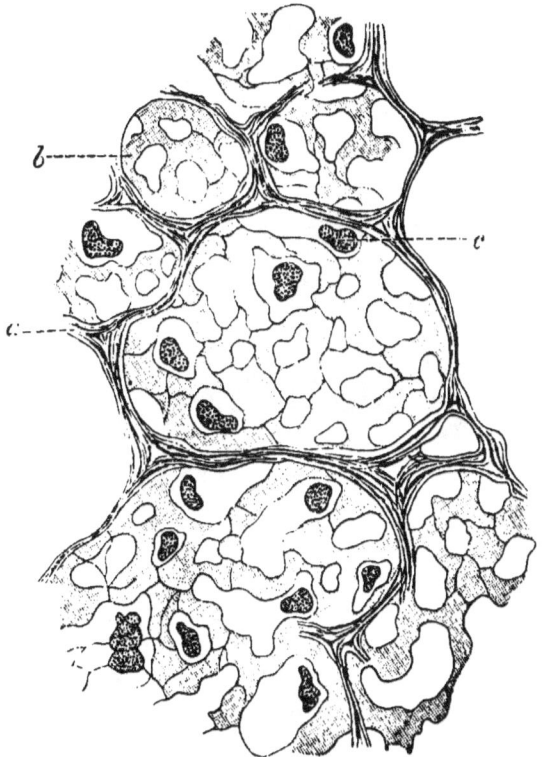

FIG. 41.—Alveolar cavities of lung of kitten. Blood-vessels injected with silver ; a, alveolar walls ; b, endothelium of injected capillaries ; c, connective tissue cells.

The inhaled particles of dust usually gain admission to the alveolar lymphatic spaces through passing, as above described, by the side of a young epithelial cell. Just as frequently, however, they first get into the cell, and are then apparently transferred by it to the underlying plasma space ; the young epithelium in this way performing a pseudo-stomatous function.

THE AIR-VESICLES.

The tissues that we shall have to deal with, as composing the alveolar wall, are therefore:— (*a*) A flat layer of epithelium more or less germinal in character according to the age of the subject; (*b*) Elastic tissue and small bundles of white fibrous tissue, with occasional connective-tissue corpuscles; (*c*) An abundant plexus of capillary blood-vessels; and (*d*) a system of plasmatic spaces communicating with the lymphatic vessels of the interlobular septa, arteries, bronchi, and deep layer of the pleura.

FIG. 42.—Section of a coal-miner's lung, showing the lymphatic spaces of the alveolar wall injected with pigment particles. × 450 diams. *a*, pseudo-stomatous openings; *b*, underlying lymphatic spaces injected with black pigment.

In many respects the tissues composing the alveolar walls are essentially those found in serous membranes. The latter are made up of a fibrous basis with connective-tissue cells, plasmatic spaces, and lymphatic vessels; and, covering the surface, there is a delicate endothelium with pseudo-stomatous openings upon it. If the tissue composing the alveolar walls were unfolded it would be hard to find any real difference between its structure and that of a serous membrane, such as the peritoneum or pleura. It will, therefore, be advisable to keep in mind, in studying the diseases of the air-vesicles, what the morbid affections of serous membranes are, and to consider in how far the causes and effects of different forms of irritation correspond in the two tissues. The only point of real difference that exists between the two, is in the origin of their endothelium; for while that of the peritoneum, and of most serous membranes, is to be looked upon as a meso-blastic structure, the pulmonary endothelium or epithelium, as we may choose to call it, is to be referred to the hypo-blast. We shall see that the distinctive characteristics implanted upon each in fœtal life are never lost after birth, for although certain morbid appearances in the one closely resemble those found in the other, yet there are some essential differences in their diseases due to the fact of their having a dissimilar embryonic origin. In one respect this is very remarkable. The endothelium covering the pulmonary air-vesicles, and that found on the peritoneum or pleura, are both liable to germinate under undue stimulation. The appearances during the germinating stage are very much the same in both, but, while the result of the germination of the peritoneal or pleural endothelium is usually the construction of fibrous tissue and adhesions, no such fibrous adhesion of the walls of the air-vesicles ever occurs as a result of a similar increased vitality. On the contrary I find, as a fact, that the resulting cells from such proliferation in the lung invariably degenerate, and that the varieties of degeneration to which they are subject are those of epithelial surfaces generally.

On the Effects of Increased Blood-pressure suddenly applied to the Blood-vessels of the Lung.

In studying the subject of acute bronchitis, we found (p. 28) that a mere sudden distension of the blood-vessels in the mucous membrane is sufficient to induce very serious changes in the bronchial epithelium. A turgescence of these, even for a few hours, causes desquamation of the superficial layers of epithelium and exudation of leucocytes into the mucous membrane. It is essential, before commencing the examination of the lung in catarrhal pneumonia, to consider what the effects of increased blood-tension are; otherwise the exact significance of catarrhal pneumonia in relation to other so-called inflammatory affections of the lung cannot be clearly demonstrated. In this preliminary inquiry I shall endeavour to show, in the first place, what the effects of suddenly increased blood-pressure in the vessels of the lung are, and, in the second place, what effects follow when it is gradually applied, and extended over a long period of time.

We all know that one of the characteristic features of what we call a croupous inflammation of a part, that is to say, where a "false membrane" is thrown out on a free surface, and where the vessels of the part are in a state of turgescence, is the suddenness of the attack. Such croupous forms of inflammation are exemplified in croup, where the mucous membrane of the larynx becomes covered with what is known as a "false membrane"; in fibrinous pleurisy, where a similar "false membrane" can be stripped off from the pleura; and in lobar or acute pneumonia, where a solid exudation of the same kind is effused into the cavities of the air-vesicles. In all these, the diagnostic points of the disease are the turgescence of the blood-vessels of the part, and the pouring out of fibrinous lymph. There is no real difference between the three, further than that of locality. In all of them, when the false membrane is removed, the underlying surface is found to be unbroken and comparatively healthy. They accordingly sometimes go by the common designation of "croupous inflammations."

What is this so-called "croupous inflammation"? How is it induced? Is it allied to "catarrhal inflammation," and, if not, what are the essential differences between them?

I shall examine these questions by reference to croupous *pneumonia* more especially, although the following observations apply equally to similar affections in other parts of the body. This lesion of the lung is very properly divided into three stages—congestion, red hepatization, and grey hepatization. In the stage of congestion the vessels are turgid with blood, but as yet there is no solid effusion into the air-vesicles, and the lung is still vesicular. When the air-vesicles are examined in this stage the whole of their capillary blood-vessels are found to be distended and engorged with blood. In some places the engorgement is greater than in others, probably from the circulation in certain of the branches, in the areas of greatest congestion, having been in a state of stagnation. The epithelium covering the air-vesicles may be observed to have desquamated in some places, and to have been cast off into the alveolar cavities. The epithelial cells can be seen rising from the alveolar wall, and, here and there, separating from it.

In the second stage, or that of red hepatization, effusion of a solid substance has taken place into the air-vesicles, rendering the lung partially or completely non-vesicular, so that it now has the consistence of liver-tissue, and will sink in water. When this solid material is microscopically examined it is found to be composed of the following elements: fibrin, blood-corpuscles of both kinds, and desquamated epithelium. The fibrinous network closely adheres to the alveolar walls, and the other elements are entangled in its meshes. In many of the air-vesicles, small hæmorrhages have evidently occurred, as lobules are occasionally noticed distended with blood. The lung in this stage is tough, and is not easily broken down. The blood-vessels at the same time are in a very turgid condition, from the large amount of blood which they contain.

In the stage of grey hepatization, the lung becomes completely solidified, has a greyish anæmic appearance, and has lost its tough consistence, being more friable than it was in that of red hepatization. This difference in consistence is due to changes which have ensued in the nature of the exudation. For, whereas, in the second stage, fibrin was its main constituent, this has now almost entirely disappeared by a process of degeneration, and a great accession to the number of leucocytes has taken place. These, in the grey stage, are almost the sole occupants of the air-vesicles, which they distend to compression, and signs of commencing fatty degeneration can be noticed in them. The turgidity of the blood-vessels has disappeared; and, although some of them may still contain blood, they have not the same engorged appearance that they had in the earlier stages.

In all other situations where croupous inflammations occur the disease goes through the same transitions, and the exudation or false membrane is made up of similar constituents, namely, fibrin, leucocytes, coloured blood-corpuscles, and desquamated endo- or epithelium.

From the study of the frog's web and the mesenteries of different animals, we know what the effect of an irritant suddenly applied to these tissues is. The arterioles spasmodically contract, and remain in a contracted state for a length of time proportional to the intensity and duration of the irritation. This is probably a reflex act, consisting in the propagation of a stimulus to a vaso-motor centre, and the giving out of increased nerve energy sufficient to keep up a more or less tonic spasm in the arterioles. We see the result of this spasmodic contraction of the arteries in the blanching of the skin which immediately follows the infliction of a burn or the application of intense cold. The death-like pallor of such parts is due to spasm of the muscular coat of the arterioles. Should the cause of irritation be soon removed, however, the spasm of the arteries ceases, the muscular coat apparently becomes wearied, and the action previously set up is now followed by an equivalent counter-action. This is characterised by the relaxation of the arterial coats, and by an increased rapidity of transmission of the blood-stream, and should no obstruction exist in the capillaries or veins, this simply produces a reddening of the part, which ceases in course of time, as the arteries regain their proper tone. No deleterious effects follow, as we can verify in the mere reddening of the hands after working in snow or other cold medium.

During the period of spasm, however, more especially

if it has been kept up for a lengthened period, the blood may have become so inspissated within the capillary vessels or veins that, when the arterial circulation returns, on the relaxation of the arterial coats, the blood-corpuscles, which have accumulated in the capillaries, will probably refuse to move onwards, and stasis, more or less complete, will follow in the circulation through the part. If this be spread over a large area, and if it is not ultimately overcome, the part dies. We see this in the effects produced by the prolonged action of cold on a part. The circulation is not re-established, the blood-corpuscles have become too firmly impacted within the capillaries, and death of the part ensues, producing what we know as a "frost-bite."

But if the obstruction of the capillaries is not universal, if it occur only here and there, then entirely different phenomena ensue, which we call "inflammatory." The part becomes painful, hot, dusky red, and swollen. When minutely examined it is found that its vessels are distended, and that there is more or less stasis at the focus of greatest irritation.

The blood-corpuscles, as shown by Addison, Zimmermann, and Cohnheim, now begin to leave the blood-vessels. The leucocytes, more especially, make their way through the walls of the veins, capillaries, and small arteries; but the coloured blood-corpuscles also escape in considerable numbers. These wander into the tissues immediately around the vessels, or, if the latter lie near a surface, as in a serous membrane, they exude on to it, and accumulate there. At the same time, it is evident that increased exudation of the *fluid* constituents of the blood has also been taking place, from the swelling seen in the part, and from the great amount of fluid contained in its lymphatic vessels and fibrous interspaces.

All these phenomena, to which we give the name of "inflammation," I look upon as *purely mechanical*, and as resulting from one cause, namely, *suddenly increased blood-pressure*. Before, however, showing what the facts and arguments are which have led to this conclusion, it will be advisable to compare the phenomena just recorded, which have been repeatedly observed in inflammation of the transparent parts of the lower animals, with what ensues in the lung of man under like circumstances, and to see whether they correspond.

The organ, from the fact of its communicating with the atmosphere, is constantly liable to irritation. Sudden changes of temperature, the inhalation of foreign bodies, and other such external agencies, tend to over-stimulate its exposed surface. We know that if a serous membrane, which it much resembles in structure, were exposed to the same vicissitudes, it would react in a most violent manner. Take, then, as an example, one of the commonest sources of irritation of the alveolar walls, namely, the sudden application of cold. We are all liable to such exposure, and no doubt the alveolar blood-vessels are constantly reacting to it. It is only in certain cases, where the parts are probably unduly stimulated, or in which the blood-vessels are in a specially favourable condition for further change, that any morbid condition of importance is induced. The first effect of the application of such stimulation, judging by analogy from what has been verified in other parts, will be to cause spasm of the branches of the pulmonary and probably also of the bronchial arteries.[1] This will last so long as the stimulation continues, and will be proportional in amount to its intensity. During this time the circulation within the capillaries connected with these constricted arteries will have stagnated, and the fluid part of their blood, probably by passing through their walls, will leave the corpuscles as the almost sole occupants of their lumina; or, as Mr. Jones has described (Guy's Hospital Reports, vol. vii.) blood-corpuscles may regurgitate into them from neighbouring blood-channels. When the period of arterial spasm is over, and when relaxation of the muscular coat of the small arteries takes place, the blood will rush

---

[1] In some unpublished experiments we have found that on the application of a current of very cold air to the lung of a frog placed under the microscope the capillaries contract to two-thirds of their former diameter.—ED. PRACT.

through them with greater freedom, but will meet with serious obstruction to its onflow in those capillaries in which stasis has occurred. The condition of many of these capillaries, we know, will entirely preclude the passage of blood-corpuscles along them, and the seriousness of the ill-effects which will result depends upon their number and on the extent of the area in which these now occluded vessels are distributed.

The main effect of this stagnation will be, that an increased strain will be thrown upon the vessels, or parts of vessels, which are still pervious, and *the tension of the blood within them will consequently be raised.* In the case of the lungs, the peculiar conditions of the circulation are specially liable to produce serious results under such circumstances, from the fact that the whole of the blood in the body must pass through them. This will certainly aggravate the effects produced by the congested stage, and may be one of the reasons why acute croupous pneumonia is so commonly met with.

The tension of the blood within the pulmonary vessels will then be suddenly raised, and the question as to what will follow, practically speaking, reduces itself to this:—What is the effect of suddenly increased pressure upon the transudation through an animal membrane when there is a liquid like blood on the compressed side of it? So long as the blood-channels are perfectly free, and so long as the venous outlets are larger than the arterial inlets, the force exerted in overcoming the inertia of the blood-column, the friction of the vascular walls, and gravity, will result in the blood rushing onwards; but if the channels through which the blood circulates are occluded at certain points, and if the same force be still applied, it is clear either that one of the three following effects will be produced, or, what is more common, that all three will be combined. It must either

(*a*) raise the tension of the blood within the vessels; or,

(*b*) it must cause an increased transudation of fluid through their walls; or,

(c) it must stretch their walls and render their porosity greater.

We know, from numerous experiments, that, if a mixture containing salts and solids of an organic nature, as, for instance, water, chloride of sodium, and albumen, be placed on one side of an animal membrane, and slight pressure be applied to it, the constituent which passes through most abundantly is the water, and that, as the pressure increases, the albumen passes through in large quantity, while the amount of salt remains almost constant under different degrees of pressure. It thus follows that if the tension of the blood be, as in the natural condition, just slightly in excess of that required to overcome friction and other hindrances, a certain amount of fluids and solids (albumins) will transude. If this pressure be raised, however, an increase in the solids of the transudation will follow; and, if it be *suddenly* raised, before the parts have had time to accommodate themselves to the new conditions, the same amount of additional pressure will have a more marked effect than if gradually applied, in causing the transudation of a fluid loaded with solids.[1]

One of the first effects, therefore, of the sudden increase of tension in the pulmonary vessels will be that a fluid rich in solids (albumins, &c.) will pass out from the blood-vessels. It will undermine the epithelium, and, as Buhl has shown, will cause desquamation of it. It will next, seeing that the walls of the air-vesicles are extremely thin and delicate, be effused into the alveolar cavities, and will there produce fibrin in amount proportional to the solids, or rather the fibrin-forming albumins which transude with it. It is important to remember that the greater part of the solid exudation of the red-hepatized stage of a croupous pneumonia is, in the first instance, poured out as a fluid. It is only when it has been effused that the fibrin, which afterwards in great part gives it its solidity, is formed. The

---

[1] Compare with this last statement the experiments of Runeberg (*Archiv. d. Heilkunde*, xviii.).

determining cause of the fibrin-formation is, in all probability, to be sought in the dead epithelium which has been shed from the alveolar walls. The occurrence of croupous pneumonia at the base, in preference to the apex, of the lung, is, possibly, in considerable part, due to the mere gravitation of the exudation, when first poured out in a fluid condition.

The presence of the fibrin in the exudation filling the air-vesicles is thus easily accounted for, if we look at the disease from a mechanical point of view; and wherever the same factors are brought into play in any part of the body, a similar fibrin-forming fluid exudes from the blood-vessels. I can see nothing more in its presence than the evidence of suddenly increased blood-tension.

I have assumed, although of course it is impossible to prove, that the real cause of the rise in blood-pressure in croupous pneumonia is that, in certain of the vessels within the lung, the blood is in a state of stagnation, so that they are, virtually, obliterated. Let us take another case, however, in which we are certain that they must have been occluded, and in which stagnation must have occurred, and see whether similar fibrinous exudation results. I have had many opportunities of studying what the effect of multiple small embolisms is upon the condition of the lung, where they are impacted in terminal branches of the pulmonary artery. It frequently happens that otherwise healthy subjects, who receive a simple fracture of a medullated bone, die from the effects of "fat embolism." The oil from the medullary cavity is absorbed and carried to the right side of the heart, and thence into the pulmonary artery; its globules cannot pass the terminal branches of the artery, and become impacted in them. The embola are minute and are present in great numbers, while they are also bland and unirritating. The state of occlusion caused by them will, therefore, be exactly that which, *à priori*, I would say should give rise to a croupous pneumonia, on the principles just discussed, namely, that certain of the blood-channels are obliterated, and that the blood forced into those which are still open is consequently at a higher pressure than under natural circumstances.

As an actual fact, croupous pneumonia, either of a general or of a localised character, is frequently associated with this embolic condition. I have in my possession several preparations of this kind taken from persons who, previous to the time of injury, were in perfect health, but who died in from forty-eight to sixty hours, with intense pneumonic effusion into one or both lungs, accompanied by widespread "fat embolism." In all respects, these lungs, with the exception of the presence of the embola in the vessels, presented the appearances of a red hepatization, and I think the conclusion inevitable, that the obliteration of the vessels was primary, the croupous exudation secondary.

It must not be supposed, however, that the *cause of the rise in blood-pressure* is always to be sought in an occlusion of certain branches of a vessel. It can easily be seen that different conditions of the blood itself would give rise to the same result, and, where these croupous inflammations are general, it is possible that this may account for them. In all cases, however, I can see nothing more in the exudation of a solid material than the action of increased, and, more especially, suddenly increased, pressure upon a fluid holding these solids in solution, and confined within an animal membrane of great delicacy.

The extra pressure existing in the lung capillaries will of itself force a thick fluid to transude into the air-vesicles, and will thus supply an excess of fibrin-forming materials. In the mechanism of respiration, however, we have another factor which may materially aid in causing inspissation of fluids effused into the lung, and which may account for the intensely solid character of the exudation in croupous pneumonia. Granted that, from suddenly increased blood-pressure, a fluid rich in solids is effused into the air-vesicles, there is a means by which this fluid may become still thicker. During inspiration the air within the air-vesicles must be at a lower pressure than that of the atmosphere, in order to effect the entrance of the latter into the chest. During expiration, however, it must be at a higher pressure than that of the atmosphere, otherwise it would not leave the chest. Its tension requires to be raised before it will rush out, and to be diminished before it will pass in at the glottis. The whole virtue of inspiration and expiration consists in the tendency of a gas to assume an equal density throughout, the sphere of greater density causing a current into that of lesser density, until the two are in equilibrium. Under ordinary circumstances this interchange from a greater to a less dense medium goes on with perfect regularity in the chest, and, in all probability, any effects which might be caused by it are counterbalanced by the greater pressure being on either side, the atmosphere or the air-vesicles, alternately. But during forced expiration with an obstructed outlet the pressure within the air-vesicles may be increased to an extent enormously greater than that of the normal state. In coughing the chest is distended with air, the glottis is closed, and the inspired air is compressed by the muscles of expiration. Its tension will now be enormously raised, and the effect of this will be, under ordinary circumstances, to cause emphysema, on the principles before explained (p. 90), when treating of emphysema as a complication of bronchitis.

If, however, a fluid containing a high percentage of solids is already present in the air-vesicles, as we have seen is the case in the commencement of a croupous pneumonia, then it is clear that a certain amount of the watery part at least of this fluid will be pressed out of the lung into its absorbents, and will leave the exudation in a still more concentrated state than it was when effused. The same principle will hold good for effusions into the pleura, where there have been coexistent forced expiratory efforts, as in coughing. If fluid has already been effused into the pleural sac, it will become inspissated by such forced expiratory efforts.[1]

[1] Dr. James drew attention to this in a most able paper on the subject of exudations and transudations, read in November 1879, before the

It is therefore clear that several causes combined will tend to render transudations into the air-vesicles, from suddenly increased blood-pressure, specially rich in solids—that is to say, in fibrin-forming constituents. It can now be easily perceived how it is that, after a few hours' illness, the whole of a lobe, or of an entire lung, may be rendered solid and non-vesicular, from the presence of fibrin and other blood-products. The exudation is loaded with solids when effused, and tends to become more and more solidified, from the positive pressure of forced expiratory efforts.

I have taken into account only one of the solid elements in the pneumonic exudation, and it may now be fairly asked, "How is the presence of blood-corpuscles to be accounted for on the mechanical principles just described? Does their presence within the air-vesicles not indicate that there is something more than mere mechanical action at work? Do they not show that the tissue has some special attraction for them?" I think it will be admitted by every one that the great mass of the cells found in the croupous exudation of acute pneumonia is composed simply of leucocytes. In the earlier stages of the disease the coloured corpuscles are quite as numerous as the colourless in the exudation, but the former very soon disappear, along with the fibrin, and are supplanted by a great increase in the number of leucocytes. In the stage of grey hepatization the number of these is enormous. They are usually accounted for on Cohnheim's theory of suppuration. The leucocytes, by virtue of their amœboid movements, are enabled to make their way through openings which naturally exist in the walls of the vessels.

Although there can be no doubt of the fact that they do exude, and that they form a large proportion of the pus

Edinburgh Medico-Chirurgical Society. (*Med. Times and Gazette*, Jan. 3, 1880.)
Dr. James, further, pointed out the fact, that the negative or diminished atmospheric pressure during inspiration will also, in the case of the pleura, aid in producing a fluid rich in solids, from the tendency to separation which will ensue between the two layers of the pleura, and the consequent aspiration produced thereby.

found in the organ when resolution occurs, yet I feel convinced that the theory of their exuding, on account of their amœboid movements, has been carried too far, and that a great deal of misapprehension exists as to the principles upon which the passage of blood corpuscles through the vascular walls is effected. Pathologists have run mad over this "exudation-of-leucocytes" theory, and I believe have mistaken its true nature. It has been supposed to afford an explanation of almost every known pathological process from the formation of fibrous tissue up to the growth and renewal of epithelium. There has been a mixing up of two processes which are often associated, namely, true histogenesis, and the wandering outwards of leucocytes; and I believe that Cohnheim and his school, although deserving of praise for bringing into more prominent light the discovery of Waller, have done quite as much harm in ignoring, in a somewhat dogmatic manner, the part played by the tissues in the processes of suppuration and repair.

It has been already stated that in the stage of red hepatization leucocytes are present in considerable numbers in the exudation within the air-cells, while, in the stage of grey hepatization, they become much more numerous, and, in fact, form the main part of the solid deposit. Their presence in the air-vesicles, under such circumstances, is quite in accordance with what we find in other "acutely inflamed" parts. All croupous "false membranes" contain leucocytes in abundance, and there can be no doubt that they somehow pass out of the vessels and accumulate on the surface of the part. It is said that the cause of their exuding is that they possess amœboid movements, by which they are enabled to insinuate themselves into the natural openings or stigmata between the endothelial cells, and gradually to push their entire body outwards. What the exact reason is for this somewhat extraordinary behaviour of the leucocytes has never been clearly explained, but it is generally assumed that the "inflamed tissue" has some special attraction for them, which induces them to leave their natural habitat. This is entirely a matter of theory to which experiment has not given the slightest support.

It certainly is a very strange phenomenon that the colourless corpuscles escape into the tissues, and the explanation offered seems insufficient to account for it. What, for instance, is to be made out of the equally well established fact that the *coloured* corpuscles, which possess no such vital movements, also exude in the same way, although in lesser numbers? This seems to be a strong argument against the vital movement theory.

The cause of the phenomenon has, as yet, therefore, never been satisfactorily explained, although it is an undoubted fact that both in the living and in the dead subject there is evidence to show that it takes place to a large extent in so-called "acutely inflamed" parts.

Before recording what experimental facts[1] I have worked out in reference to the cause of this diapedesis of leucocytes through the walls of the blood-vessels, allow me to call the reader's attention to the following points, which are of primary significance in understanding it:—

(a) *The blood-vessels of the part in which the exudation of leucocytes occurs are in a state of acute distension.*

(b) *The circulation in many of them is in a condition of stagnation, and it is within those in which stagnation is greatest that the exudation of leucocytes is most abundantly observed.*

(c) *The exudation of leucocytes is accompanied by swelling of the part, from infiltration of fluid into its fibrous interspaces. This fluid is accounted for by increased transudation from the blood-vessels, indicating an increased blood-pressure. Ligature of the vein leading from the limb of an animal produces the same effects.*

(d) *It is where the return of blood is suddenly arrested that these exudative phenomena are observed.*

In the natural endothelial lining of blood-vessels more especially of veins and capillaries, there are undoubted

---

[1] An abstract of these experiments is given in the *Proceedings* of the Royal Society of Edinburgh, Session 1881-82, p. 370.

openings or deficiencies, through which blood-corpuscles might pass if the necessary forces were brought into play. These forces must be situated either in the corpuscles themselves, or must be extraneous to them. If they are extraneous, then it is clear that in the normal condition of the circulation they do not come into play, or are diverted, so that the corpuscles are driven in the direction of the blood-stream, instead of through the walls of the blood-vessels. If they are contained within the corpuscles themselves, then it is just as evident that they are not exerted, or that they are so counteracted, in the natural circulation of the blood, that few, if any, of the corpuscles manage to escape from the lumen of the vessel.

Putting aside the consideration of any forces that might be supposed to exist within the blood-corpuscles themselves, let us examine the effect of extraneous forces alone in causing their diapedesis. The following experiments, of a purely mechanical nature, seem to have a very important bearing on this subject: I made a solution of gelatine, glycerine, and water, of such consistence that when cold it formed a soft jelly. This was cut into pieces about the size of a small pea, and these were placed, suspended in water, upon one side of a dialyzer. The dialyzer was then connected with a column of water about three feet high, and, with its contents, was immersed in an outer vessel of water. There was now contained within the dialyzer a mixture of gelatine masses and water, which we may, for the sake of illustration, suppose to represent the blood with its corpuscles. This was subjected to the pressure of a column of water three feet high, the dialyzing membrane being sufficiently strong to prevent rupture. We may further suppose the whole apparatus to represent a blood-vessel in which the circulation has been suddenly arrested at some part of its course, and in which the blood has come to a standstill. The blood-pressure formerly exerted in propelling the blood through the vessel would thus be diverted against its walls, and, as we saw before, would cause the transudation of more fluid, and of a fluid containing more solids, than when the circulation was free. It would also have the effect of stretching the walls of the vessel, and, of course, if this occurred suddenly, it would tend to enlarge all the natural endothelial openings, and also to separate the one endothelial cell from the other. What effect will be produced upon the blood-corpuscles?

Let the dialyzing membrane now be perforated with a needle, so as to imitate the stigmata and other endothelial openings which exist in the distended blood-vessels. These small openings are out of all proportional size to the large masses of gelatine contained within the dialyzer. They are mere needle-point apertures, while the masses of gelatine, as before mentioned, are about the size of a small pea.

The moment an aperture is made a little capillary jet of water is ejected, but this is almost immediately closed by a mass of gelatine. Many such openings may be made and the same thing occurs in all of them; a mass of gelatine becomes applied to the perforation and stops the stream of water which is issuing from it. In the course of half a minute or longer, according to circumstances, an attenuated process of the gelatine mass is found to have made its way through the membrane, and projects on the outer side of it in a little bead-like protrusion. This increases in size, and, finally, the whole mass makes its way through, and is set free into the outer vessel of water in which the dialyzer is suspended. The passage of a mass of gelatine from the one side of a dialyzing membrane to the other, through an aperture probably not one-fiftieth part of its own diameter, can thus be accomplished, merely on account of the tension of the fluid within the dialyzer being greater than that outside.

A very strong argument in favour of a purely mechanical and extraneous cause being the agent of expulsion of the blood-corpuscles from the vessels in acute inflammation seems to be established by this fact; for not only is a solid substance such as gelatine capable of being transmitted through a membrane in this way, but the whole of its transformations in shape are so like those of a blood-corpuscle, in being extruded from a vessel, as to render the comparison all the more striking.

Let us now alter the conditions of the experiment in some particulars. Instead of a dialyzer, let us employ a glass tube with a number of minute perforations in it, and let the pieces of gelatine be again suspended in water within it. Let both ends of the tube be open, and let the same pressure as before be applied to one of them. The fluid then circulates freely through the tube and issues in jets from the holes in it, as well as in a free stream from the open end. Let us now watch the pieces of gelatine circulating in the tube and see how they behave in their passage along it. Some of them, which are at the periphery of the tube, adhere to and are projected through its openings, but by far the greater number remain in the centre, and move along at an equable pace in a continuous stream. The amount of diapedesis which takes place in this experiment is infinitely less than where the stream is arrested.

From these observations we learn two things:—First, why the leucocytes exude in greater numbers than the coloured blood-corpuscles; and second, why it is that, in the natural circulation in a part, only a small number of blood-corpuscles make their way out of the blood-vessels. It is a fact, which can be verified in the study of the frog's web, that while the hæmatocytes move in a rapid stream in the centre of the vessel, the leucocytes adhere to the walls and go at a much slower pace. We can easily understand, therefore, why it is that they should be exuded in such large numbers. As they roll along, in close contact with the endothelial lining of the vessel, they will naturally be applied over the openings which exist in it, and will more readily than the corpuscles in the centre of the stream be pushed outwards by the pressure of the blood from behind. When stagnation occurs at a point in front, of course the *vis e tergo* will be wholly diverted against the wall of the vessel, and will vastly increase the tendency to diapedesis. In the natural circulation of a part, where there is no obstruction in front, the *vis e tergo* is, in great part, expended in driving the blood along the natural channels, and the corpuscles, being carried forwards in the current, have little tendency to exude. It is only in the case of certain of the leucocytes, which, in moving along more lazily, in close contact with the endothelium, become applied to a stigma in the wall of the vessel, that any tendency to exudation occurs. The cause of the corpuscles not exuding in any great number, where the natural channels are quite pervious, is very much the same as that which enables a locomotive to run on a straight line instead of diverging into other paths. The *vis e tergo* is exerted in the direction of least resistance. Place some obstacle in that path, and then, as in the case of the blood-corpuscles, the onward progress of the locomotive is turned aside, and it is thrown off the line. Exactly in the same manner, the blood-corpuscles, so long as there is a free path in front, will run along this rather than out at the narrow stigmatous apertures. Obstruct that path, however, in front, either completely or partially, and let the *vis e tergo* be the same, and then, instead of passing onwards, the only course left for them to pursue is through the walls. If, as has been proved, there exist here natural openings, and if these openings be stretched, then the blood-pressure will drive the blood-corpuscles and the other blood-constituents through these, just as in a dialyzing membrane. If, further, the vital movements of the leucocytes come into play, they may aid in their extrusion, but that these are the primary cause of their expulsion I cannot believe, seeing that diapedesis occurs mostly under circumstances where undue pressure is exerted upon the vascular walls, and also seeing that a similar phenomenon can be produced with bodies composed of dead material such as gelatine.

It is therefore clear that, quite independently of any vital properties, the whole constituents of the blood, solid and fluid, can, under circumstances of undue pressure, pass through the walls of the blood-vessels into neighbouring parts.

The entire process, I feel persuaded, from what can be seen histologically in such parts, and from experimental evidence, is purely mechanical.

Another strong argument in favour of this croupous pneumonia and other forms of croupous inflammation being merely manifestations of undue blood-pressure suddenly applied, is seen in what can be effected in the way of treatment of such cases. Notwithstanding all that has been said to the contrary, and in spite of the prevailing *fashion* of the present day, I believe that venesection is the one sovereign remedy in this disease. To any one who has seen, as I myself repeatedly have, the instantaneous relief afforded by the abstraction of blood in this disease, even in cases where the constitution of the patient would be considered unsuited for the operation, the conviction is irresistible, that physicians have let a practice of the utmost value fall into disuse. It is not the object of the present series of papers to enter into the therapeutics of the subject, but, in passing, I cannot refrain from expressing my strong conviction that in venesection we have the one means of lowering the blood-pressure and cutting short the disease. Once the undue blood-pressure has been relieved, the exudation of its solid constituents must cease, and time will thus be afforded for the circulation in the part to recover itself. The hard wiry pulse of a person suffering from croupous pneumonia simply expresses the high tension of the blood. Remove part of the latter, and relieve this tension, and you will cut short the disease. The great mistake which has been made in the practice of venesection, is that of employing it in the wrong kind of pneumonia—that is to say, in instances of catarrhal pneumonia. The main object I have had in view in this digression into the subject of croupous pneumonia is to show the essential difference between these two diseases, which both, quite wrongly, go by the name of "pneumonia." It will be shown in the sequel, that while acute or croupous pneumonia is simply a manifestation of suddenly increased mechanical pressure upon the blood-stream, catarrhal pneumonia is of a different nature, and that the indications for the treatment of the one are opposed to those of the other.

It is premature, at present, before I have shown what catarrhal pneumonia is, to make any deductions as to its essential nature, but having gone into the subject of the essential nature of croupous pneumonia, we are now prepared to ask the following question, the correct answering of which is of the utmost utilitarian value: "Is what we call a croupous pneumonia to be reckoned as an inflammation?" This of course will depend upon what we mean by the term, but if it be applied to the exudation, into a tissue, of a substance composed of the normal constituents of the blood, can we, with any logical reasoning, call this by the same name as a lesion in which the essential of the morbid process is the undue stimulation and proliferation of the natural elements of the tissue? We shall see that catarrhal pneumonia belongs to the latter class of diseases, and, to me, it seems the height of absurdity to call this condition by the same name as that in which there is mere effusion of the blood-constituents. There could be nothing more misleading; and until physicians come to recognise the difference in the pathology of the two processes, their treatment will be a matter of mere empiricism.

On the Effects of Long-continued Excessive Blood-pressure in the Lung.

Such then are the results of a sudden rise in blood-tension in the vessels of the lung, due to an impediment to the onward flow of blood in certain branches of the pulmonary artery, its capillaries, or the pulmonary veins. The constituents of the blood exude in greater or less quantity, and come to fill the air-vesicles, producing what is commonly known as a "croupous pneumonia." It will next be necessary to examine what changes occur in the organ as a result of increased blood-pressure *gradually* applied and continued over a long period of time.

The best opportunity for studying this is afforded in cases of chronic mitral disease, where the blood is retarded in its flow within the pulmonary vessels, either from the constriction or incompetence of the mitral orifice. When such a lesion is suddenly produced croupous pneumonia and pleurisy are frequently set up by it. It is well known that recent valvular lesions, from the acute endocarditis of rheumatic fever, are often followed by pneumonia or pleurisy. The two are not generally looked upon as cause and effect, but there are good grounds for believing that such a relationship exists between them. The endocardium composing the valve becomes injured first, this trammels the pulmonary circulation to a corresponding degree, and, as the injury of the valve is suddenly produced, a rapid rise in the blood-tension ensues in the lung, causing increased transudation of the solids of the blood, as before explained. That the croupous pneumonia, croupous pleurisy, endocarditis, and swellings of the joints, are all manifestations of the same disease, as is generally supposed, seems improbable, for while the pneumonia and pleurisy essentially consist in the pouring out of the solids of the blood, the others—the endocarditis and affections of the joints—are true hyperplasiæ of connective tissues, in which blood-vessels are either absent or are present in particularly small numbers, but in which connective-tissue elements are abundant. I should therefore look upon the acute pneumonia and pleurisy of rheumatic fever as, in most cases, effects of the sudden interference with the pulmonary circulation, from a valvular lesion of the heart.

We all know, however, if the patient recovers from the primary effects of such a valvular insufficiency, and if the balance of the circulation has, in a measure, been restored, by the parts accommodating themselves to the new circumstances, that pulmonary symptoms more or less severe are constantly liable to arise, from the permanent injury which the valve has suffered. Hæmoptysis is frequent, and symptoms indicative of bronchial œdema are of common occurrence. Both of these effects are due to the organic deficiency in the mitral or other valve.

When the lung of a person who has suffered from such a valvular lesion is examined after death, it is found to be in the morbid state known as "brown induration." The pleural vessels are deeply congested, wedge-shaped hæmorrhages are found at its edges, and sometimes in the centre of the organ, and throughout its substance indurated patches with ill-defined borders are scattered at intervals. These indurated patches are almost like little pneumonic deposits, are only partially vesicular, and have a brownish-red colour. They are best seen in the centre of the organ, in the vicinity of the larger bronchi. The organ is also usually more or less œdematous.

When these brown and indurated patches are microscopically examined they are seen to have the appearance shown in Fig. 43, which represents one entire alveolus, with portions of several others. An individual patch is made up of several lobules, whose air-vesicles have the appearance of that in the figure. The capillaries (a) on the alveolar walls are distended and engorged with blood to an extreme degree. Each branch rises from the alveolar wall in the shape of a loop, and projects into the alveolar cavity for a considerable distance. The extent of this capillary dilatation throughout the lung may be conjectured from what is seen of it in a single air-vesicle. The whole of the alveolar capillaries are in this condition, but those which are near a bronchus generally exhibit the greatest dilatation. The branches of the pulmonary veins and artery are also much distended, but not in so marked a manner as their capillaries.

The air-vesicles within the patches above referred to show some solid contents. They are represented in the figure. Blood pigment of a dark brown colour is of constant occurrence. It lies free either in the alveolar cavity or in the alveolar wall, or it is contained in large cells (b). These cells are flat, and possess one or more nuclei. They are the epithelial cells of the alveolar wall which have desquamated, and which have been cast into the alveolar cavity. At the lower part of the figure one of these is seen on transverse section in the act of being removed from its attachment. Besides these, blood-corpuscles of both kinds (c) are common, the leucocytes absorbing some of the brown blood-pigment. Groups of three or four air-vesicles are sometimes seen in such a patch into which hæmorrhage has occurred, part of the pigment above described being evidently derived from such extravasations.

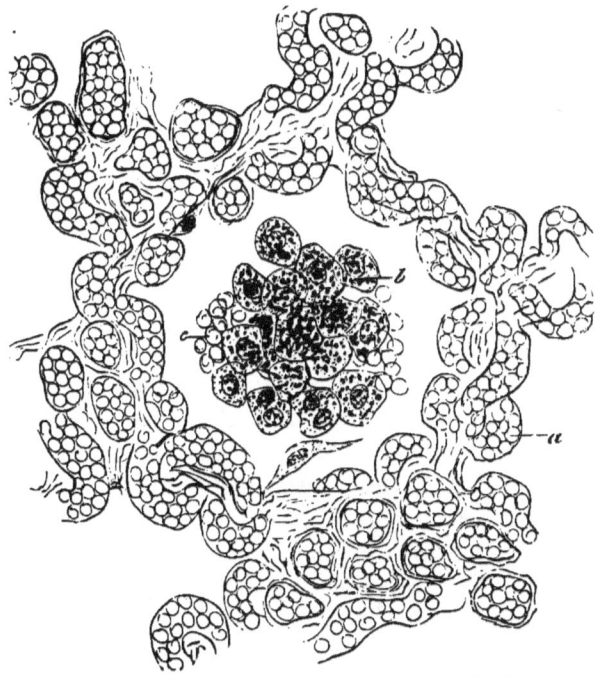

Fig. 43.—Section of an alveolus from lung of a person who died from mitral disease. *a*, distended and projecting alveolar capillaries; *b*, desquamated epithelium; *c*, blood-corpuscles extravasated into alveolar cavity.

It occasionally happens that, within certain air-vesicles, fibrin, such as is found in a red hepatization, is seen, but this is unusual.

All these appearances are due to one agent, namely, long-continued increased blood-pressure. The first effect of this is to cause widening and engorgement of the alveolar capillaries. Serum exudes from these, and carries with it more or less blood-pigment. The serous exudation undermines the alveolar epithelium, and loosens it, so that it is cast off into the cavity of the air-vesicle; while the cells, either before or after desquamation, absorb some of the blood-pigment, and incorporate it with their nuclei or protoplasm. From the distended state of the vessels a few leucocytes manage to escape into the alveolar cavities, but the greater number of blood-corpuscles found in the air-vesicles are poured into them as a result of extravasation. When the distended condition of the capillaries is taken into account, it cannot be wondered at that rupture of them should constantly be liable to occur, and indeed it is remarkable that rupture is not more frequent.

The "brown induration" therefore is merely a mechanical product, and closely resembles the lesion which we saw in the walls of the bronchi under like circumstances. The distended vessels keep up a constant state of œdema of the part, which totally unfits it for the growth of proper epithelium. The cells are rejected as soon as they are formed; and it is these, along with the small hæmorrhages, which give rise to the feeling of induration in the brown-coloured patches. These epithelial cells, however, are comparatively scanty, and do not seem to proliferate to any great extent after desquamation.

The effects of increased blood-pressure upon the pulmonary vessels are therefore different according as it is applied suddenly, or gradually and continued for a lengthened period. In the former a croupous exudation is thrown out; in the latter, the capillaries become chronically distended, the epithelium desquamates, and a certain amount of serum exudes. Why is it that such a difference in the effects should thus result?

If the delicate capillary or venous wall, composed of or covered by an endothelium having natural openings or deficiencies, be suddenly stretched, a greater widening of these openings will result than in the case where the distension is of a chronic nature. The cause of this is apparent. Sudden distension of the vessel will tend to separate the one endothelial plate from the other, ending, as it frequently does, in complete rupture; but even when the latter result does not ensue, the mere stretching of the vessel will widen the natural openings between the endothelial cells, and will leave so many apertures through which the blood may escape. All these deficiencies would be repaired in course of time, and it is not hard to see that the sudden distension of the vessel is one reason why blood-corpuscles escape when there is a rapid increase of blood-pressure, and why the tendency to such is lessened, although not necessarily prevented, by its long continuance. There is, however, another reason why the blood-corpuscles do not exude in great numbers in instances of chronic regurgitant pressure from valvular lesion, namely, that there is no actual stagnation of the blood in the capillaries, but merely a retardation of its onward progress. The stream of blood passing through them, although perhaps slower in its motion, is nevertheless continuous, and hence the blood-corpuscles will tend to be swept along the natural channels rather than out at the endothelial openings in the wall. Very different is the case in acute inflammation in which there is an obstruction at one point in front, and where the whole energy of the circulatory apparatus comprised in the *vis e tergo* is consequently exerted against the vascular wall. In such a case the corpuscles are forced to pass out at the lateral openings of the vessel, on account of the blood-current being directed into these.

But in cases of continuous increased blood-pressure it must also be remembered that the fibrin-forming albumins do not exude in such great quantity as where it is suddenly applied, and the reason of this is in part the same as that which we have just given to account for the small number of blood-corpuscles in the alveolar exudation of mitral disease. The wall of the vessel is less porous, and prohibits the passage of the liquor sanguinis in great quantity. Runeberg (*loc. cit.*), moreover, has shown that solutions of albumin, subjected to pressure within an animal membrane, transude in much greater quantity when there is a sudden accession of pressure than when the pressure is continuous and of constant amount. We can easily understand how this will apply to the case in point, where the increased pressure is continued, it may be, for many years, and where a comparatively small amount of fibrin-forming albumins transudes. It is undoubtedly a sudden stretching of the vascular walls which is the cause of the rapid exudation of the blood-solids, widening, as it must, all the pores of the tissues composing the vessel, and allowing of a freer transudation.

In addition to these two factors we must bear in mind that in a chronic lesion the lymphatics have had time to accommodate themselves to the new functions required of them in removing superfluous exudation, whereas in the acute disease they are suddenly called upon to discharge such functions without being prepared for the task. They become choked with fibrin so as to be temporarily useless, and the exudation still continuing to be poured out from the vessels must of necessity accumulate in the part and be forced on to its free surface. It is not until this fibrinous accumulation in the lymphatics is removed by degenerating that they are again called into play.

There is therefore no scientific difference in the determining cause of dropsy, brown induration, and croupous inflammation of the lung further than one of degree. All three are the results of the same agent differently applied. This agent is the blood-pressure, and, from whatever cause arising, if it be increased, one or other of the three lesions will result. If suddenly increased, not only the albumins of the blood, but also the corpuscles pass into the tissues; and from these, fibrinous lymph, or what is the same thing, blood-clot, is constructed. If it be gradually increased, then the effects are either simple œdema of the organ, or brown induration, according to the duration and degree of the pressure. I know of no relationship more important in pathology as applied to medicine than that of the effects on a part of different degrees of blood-pressure; and, in the present inquiry, if we clearly keep in view the points specially referred to in connection with this subject, it will be of the greatest aid in appreciating the true nature of catarrhal pneumonia.

CATARRHAL PNEUMONIA: FIRST STAGE.

The course of the disease which we are now about to consider can, with the greatest justice, be divided into three stages, which correspond with three distinct phases in its clinical history. The first I call the *acute or sub-acute stage;* the second that of *caseation;* and the third that of *phthisis,* or *destruction of the lung.*

The first, acute or sub-acute stage, commences with what the patient usually describes—and very correctly—as "catching a cold." It is the stage of catarrh of the air-vesicles. That is to say, he has been exposed to some vicissitude of temperature, and suffers shortly afterwards from cough and mucous expectoration. The symptoms are primarily those of bronchitis, the pneumonic phenomena being *gradually* superadded. Adults seldom die in this stage, recovery or a lapse into a chronic condition being the usual course which the disease pursues. In children, on the other hand, especially after measles or whooping-cough, it frequently proves fatal.

The lung, after death, presents the following appearances:— There is an absence of acute pleurisy, and adhesions of any kind between the pleural surfaces are rare. In this respect the disease forms a marked contrast with croupous pneumonia, in which fibrinous pleurisy is an almost constant accompaniment. The organ, when removed from the chest, feels vesicular throughout, often more so than a normal lung, from the difficulty which the air experiences in leaving it. On the surface, however, isolated lobules or groups of lobules are seen, having a leaden or purple colour and which are almost totally non-vesicular, while the adjacent parts of the lung are more vesicular than usual, amounting, in some cases, to absolute emphysema. Solidification of the lung, as in croupous pneumonia, cannot be perceived on grasping it in the hand, and portions of it cut off do not sink in water.

The mucous membrane of the bronchi is always much congested, and from the bronchial openings more or less greyish-yellow mucous secretion can be expressed. The lung contains a medium amount of blood, and, when exposed for a short time to the influence of the atmosphere, becomes of a bright scarlet colour. Over its cut surface, more especially towards the periphery, are seen irregularly-shaped pneumonic patches, corresponding in size to that of a lobule of the lung, which have a greenish-yellow colour, and from which, when squeezed, a little mass of yellow catarrhal fluid, like that contained in the bronchi, can be pressed out. Those patches which are adjacent to the pleura sometimes have a wedge shape. In all of them the border is extremely indefinite, and they are soft, somewhat raised, and slightly vesicular. The lung feels, when the hand is passed over it, like a mass of frog's spawn. On account of the solidification being confined to a lobule, the name of "lobular" is sometimes applied to this form of pneumonia, in contrast to that of "lobar" given to the croupous variety.

When such a catarrhal pneumonic patch is microscopically examined, with a magnifying power of about fifty diameters, it has the appearance represented in Fig. 44. In the centre of the patch there usually is a small bronchus (b), more or less distended with cellular bronchitic products, while the remainder of the patch is made up of a group of air-vesicles surrounding the bronchus, loosely packed with catarrhal cellular products. It is these catarrhal products which can be squeezed out in the fresh state from the pneumonic patch, and which give rise to the partial consolidation. There is not usually any fibrin in the exudation, so that the pneumonic patch is never so tough as in the case of croupous pneumonia, in which, at a corresponding period of the disease, fibrin is the chief cause of the solidification. The mucus, which is largely present in the catarrhal pneumonic effusion in this stage, gives it the consistence of a viscid fluid.

The alveolar contents at this time consist of two elements —cells and a mucous fluid. Discarding the fluid part of the secretion for the present, let us more particularly examine the cellular elements. A group of these cells is shown in Fig. 45, where it will be seen at a glance that the members of the group differ in size and general contour. The form most commonly observed is represented at a. It is a large, flat body with finely-granular protoplasm and usually two or

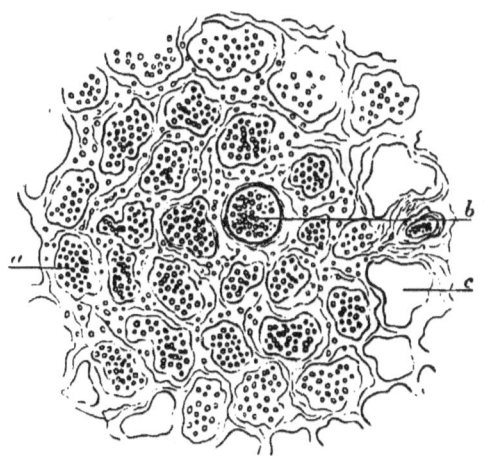

FIG. 44.—Acute catarrhal pneumonia. Group of air-vesicles and small bronchus infiltrated with catarrhal effusion. × 50 diams. a, infiltrated air-vesicle; b, small bronchus, also infiltrated; c, an empty air-vesicle at the periphery of the pneumonic patch.

more nuclei. There is evidence, in the occasional dumb-bell shape of the nucleus, that division and multiplication have been going on. The difference in size which the nuclei frequently show (a) supports this idea. Certain of these cells, however, do not show any nucleus, but, on the contrary, exhibit undoubted evidence of fatty degeneration (b). Oil globules are visible in them, at first few in number, but, subsequently, converting the whole cell into a compound granular corpuscle (Fig. 48, *a*). Other cells of smaller size (*c*) are also abundantly found, each having a large nucleus with delicately-granular protoplasm. These also show clear evidence of dividing—the nucleus first, the protoplasm afterwards. Some of them are occasionally seen to be fatty. A few bodies of round shape, evidently blood leucocytes, are sometimes met with, but not often, and they do not form an essential element of the catarrhal secretion. Small hæmorrhages into the deep layer of the pleura, or into the adjacent air-vesicles, are occasionally present; but they are of local occurrence, and are never in anything like the abundance found throughout the whole lung in croupous pneumonia.

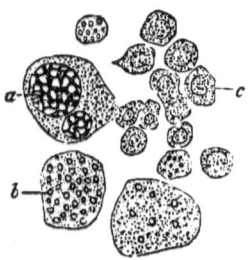

Fig. 45.—Catarrhal cells. Acute catarrhal pneumonia. × 450 Diams. *a*, large catarrhal cell with two nuclei ; *b*, an epithelial plate becoming fatty ; *c*, germinating catarrhal cells.

The origin of such catarrhal cells is apparent when the alveolar wall is carefully examined. In the description of the natural epithelium covering the walls of the air-vesicles it was shown (Figs. 39 and 40) that although the greater number of the epithelial cells were flat scales, like those seen in an endothelium, there were groups of smaller cells constantly met with among these which were more germinal in character, and which were evidently in a state of active proliferation. In acute catarrhal pneumonia the germination noticed in these groups of young epithelial cells is vastly increased, so that, instead of being scattered here and there over the alveolar surface, they entirely cover it. The older fully-formed cells are cast off and soon become fatty, constituting the fatty cell-plates represented in Fig. 45 at *b*, while their place is taken by cells of an embryonic character, and much smaller in size. A drawing of the appearance presented by the alveolar wall in this stage of catarrhal pneumonia is given in Fig. 46. So far as could be learned, the child from whose lung the drawing was made had suffered from acute catarrhal pneumonic symptoms for a few days. The reader is supposed to be looking into the interior of the air-vesicle, the part indicated by the letter *b*, corresponding to a surface view of the alveolar wall, while the letter *a* indicates the same cut transversely. A profile view of the alveolar epithelium is also thus obtained at the letter *c*. The epithelium (*b* and *c*) lining the air-vesicle can now be distinctly seen, even although it was not stained with silver; the reason of this being that the cells have lost the character of endothelial scales, and have now become much more prominent objects. At the same time their protoplasm has assumed so granular a consistence that the position and borders of the cells are better defined. Instead of there being, as in the natural state, groups of germinal epithelial cells here and there, the whole alveolar wall is now covered by them. They are, however, even more germinal in character than those normally present, and, when seen on section (*c*), are noticed to project for a considerable distance into the alveolar cavity. They are all highly nucleated, the nuclei being proportionately large compared with the protoplasm. Some of them have two, others three or four nuclei, and active division of these can be easily observed.

New cells are thus constantly being produced at a much greater rate than is necessary for the mere investment of the alveolar wall; and a large number, being unused for this purpose, are thrown off into the alveolar cavity as waste products. It is these which constitute the bulk of the cells found in the catarrhal-pneumonic accumulation. When the cells are being cast off they are seen first to rise above the surface of the alveolar wall (Fig. 46, *c*); the attachment of the cell then becomes more and more attenuated, until a

FIRST STAGE. 141

pyriform-like cell is produced, as at *d*. The delicate stalk which still attaches this to the alveolar wall at length gives way, and then the cell is set free and is thrown off into the alveolar cavity.

The source of all these new cells is to be found in the germinal groups present in the natural epithelial lining. The older fully-formed epithelial plates are cast off, and, although

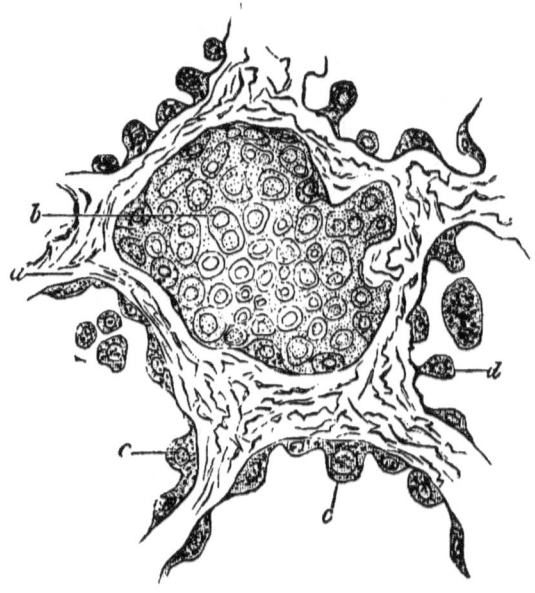

Fig. 46.—Acute catarrhal pneumonia. Surface view of the wall of an air-vesicle. *a*, transverse section of alveolar wall; *b*, alveolar cavity, showing the alveolar wall covered by germinating epithelium; *c*, germinating epithelium seen on profile; *d*, one of these germinating cells becoming separated from the alveolar wall. × 400 diams.

they also accumulate in the alveolar cavities, take no further part in the catarrhal process.

If this description be compared with that previously given of the catarrhal changes in the bronchi, it will be seen that the two exactly correspond. For, as the deep or germinal layer of the bronchial epithelium was that from which the catarrhal cells were thrown off, so here, in the alveolar cavities, it is from germinal epithelial structures of the same nature that the pneumonic elements are produced. In both cases the fully-formed epithelial cells are primarily rejected, and take no part in the germination, while, in each instance, the catarrhal process is a mere exaggeration of that which occurs in the natural epithelial repair.

When these germinal and other epithelial cells have been cast off, they accumulate in the air-vesicles, and, mixing with a little mucus, form what we understand as the catarrhal-pneumonic effusion. The mucus which this contains renders it viscid, and hence it tends to adhere to the alveolar walls, and can with difficulty be expelled from the alveolar cavities by expiratory efforts. Inspiratory efforts draw it outwards towards the pleura, so that the infiltrated air-vesicles are more numerous towards the periphery than at the centre of the lung.

The appearance which the air-vesicles present when distended with catarrhal secretion is given in Fig. 47, taken from a lung in which the blood-vessels were artificially injected. Each alveolar cavity contains a mass of cellular epithelial products closely adhering together by means of the mucous fluid in which they are suspended. The mucus is liable to become precipitated under certain circumstances, and then gives rise to a granular appearance. The epithelial cells, after being shed, have the power of maintaining an independent existence for some days, but sooner or later all of them begin to show indications that their vitality is destroyed. The first sign of this is noticed in the commencement of fatty degeneration. Oil globules appear in their nuclei or protoplasm, and before long the whole cell becomes converted into a compound granular corpuscle (Fig. 48, a).

Why there should be such variety in the character of the cells met with in the contents of the air-vesicles in catarrhal pneumonia can now be easily understood. The largest cells, which are flat and devoid of nuclei, are the desquamated epithelial plates, the smaller nucleated bodies are the embryonal progenitors of the same, while the compound granular corpuscles are either of these in a state of fatty degeneration.

The surrounding *capillary* blood-vessels of the alveolar walls I have not found to be markedly congested in acute catarrhal pneumonia. The small arteries and veins usually contain a considerable amount of blood, but not in any very great excess of that which is seen in many normal lungs. There is not any evidence of any widespread stasis in the

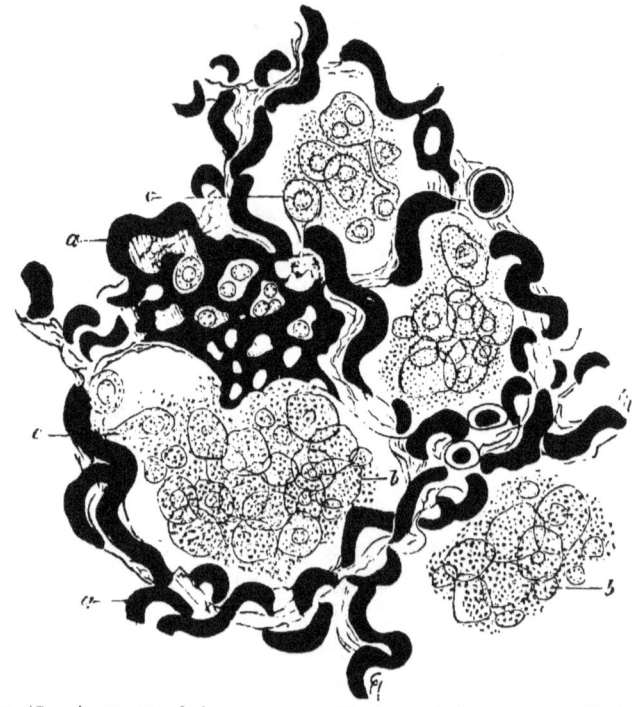

Fig. 47.—Acute catarrhal pneumonia. Blood-vessels injected. × 450 diams. *a*, injected capillaries of alveolar wall; *b*, catarrhal cells lying in the alveolar cavities; *c*, the same, sprouting from the alveolar wall.

alveolar capillaries, as in croupous pneumonia. The absence of fibrin or other blood-products in the alveolar contents shows that there has not been any sudden rise in blood-pressure.

The two diseases, croupous and catarrhal pneumonia in their acute stages, are, therefore, totally different in their nature; for, while the former is characterised by the exudation of the solids of the blood into the air-vesicles, the latter is essentially an epithelial proliferation. Both are probably caused by the same agency, namely, an undue amount of stimulation of the alveolar surface; and the difference in the result of this stimulation is, in all probability, dependent upon the relative strength of the circulating apparatus. In what are known as "full-blooded" individuals, with powerful hearts and a free circulation, I would expect that the reflex spasm of the arteries, and stasis of the first stage of irritation, would be followed by a much greater re-action, and a greater rise in blood-pressure, than in persons of a weak constitution and possessing a feeble propelling power of the heart. In the latter class of individuals, instead of the primary stasis being followed by a great increase of blood-pressure, the amount of blood and the propelling power of the heart are insufficient to suddenly force out the solid blood-constituents. The undue stimulation accordingly expends itself merely in irritating the embryonic, sentient, epithelial cells of the alveolar wall, and causes an increase in their function. The comparatively sub-acute course of a catarrhal pneumonia, as compared with the sudden onset of the croupous variety, coincides with this view of the two diseases; for, while the solid blood-constituents may be poured out in such quantity as to infiltrate a whole lung in a few hours, the catarrhal changes proceed much more gradually, and are usually limited to certain districts, where evidently the irritation has acted most strongly. If we produce a peritonitis—which closely resembles catarrhal pneumonia—artificially in an animal or in man, the focus of greatest proliferation is always to be found near the point of irritation   In the same way, we can easily understand that in an acute catarrhal pneumonia there may be certain groups of air-vesicles which are more stimulated than others, and in which consequently the epithelial proliferation will be most evident. In the case of a croupous pneumonia it is different, for if, in a certain lobe, or throughout a whole lung, there be vessels which are suddenly rendered impermeable in front, the whole of the blood-column entering

FIRST STAGE.

the portions of these vessels which are still pervious will be under a greatly increased pressure, and the exudation of the blood-constituents will consequently tend to become general. It can therefore be perceived how a catarrhal pneumonia should be lobular, and a croupous lobar. The infiltration in the one is due to a local cause, but in the other it is dependent on a cause acting over a wide surface.

The catarrhal cells having accumulated within certain groups of air-vesicles, they soon all become more or less fatty

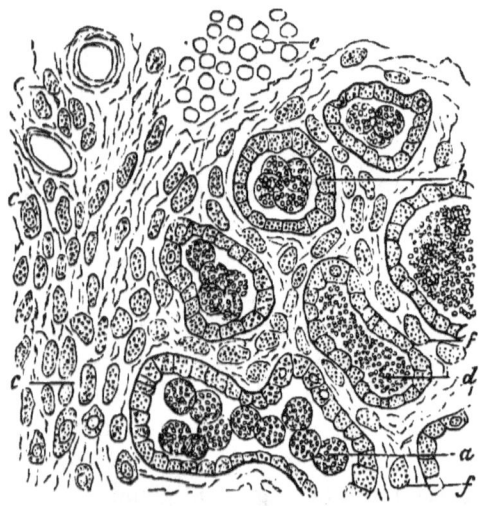

FIG. 48.—Acute catarrhal pneumonia, showing a thickened lobular septum and the contents of the air-vesicles. × 400 diams. *a*, compound granular corpuscles in air-vesicles; *b*, germinating alveolar epithelium; *c, c, c*, side of a thickened lobular septum running out to the pleura; *d*, oily emulsion formed by destruction of alveolar contents; *e*, a small hæmorrhage into deep layer of pleura; *f, f*, cellular infiltration of alveolar walls.

(Fig. 48, *a*); and, according as this fatty degeneration is of a moist or of a dry nature, depends the future history of the case. If they undergo a moist fatty degeneration, that is to say, if serous fluid is abundantly mixed with their fatty *débris*, resolution can and probably will occur, from absorption or expectoration of the oily and albuminous products of the cell-destruction. If, however, the fatty degeneration which

L

the catarrhal cells undergo be of a dry, or, as it is called, caseous nature, then absorption is impossible, and the accumulated cell-products now lie in the lung-tissue as foreign bodies, and induce what I call the second stage of the disease.

The difference in the character of the fatty degeneration which they undergo depends probably on several factors, to be more fully explained afterwards, but chiefly on one, namely, the amount of serous fluid which transudes from the vessels. The more acute the attack of catarrhal pneumonia is, and the higher the blood-pressure, within certain limits, the more probable is it that resolution will occur in the first stage. The more insidious the onset of the first stage, the greater likelihood is there of the catarrhal products drying in the air-vesicles and undergoing caseation. The difference in the two processes, dry and moist fatty degeneration, makes all the difference between rapid recovery and the passage into a phthisis pulmonalis. Could we artificially induce, in this first stage, a slight œdema of the lung, there would be very little danger of caseation occurring. It is a remarkable and significant fact, as bearing on this, that *individuals who suffer from mitral lesion of the heart, and whose lungs are always more or less œdematous, never, so far as my experience goes, die from caseous catarrhal pneumonia supervening upon this.* The reason is apparent, for although catarrhal changes are by no means infrequent in such lungs, the catarrhal products, when degenerating, mix with so much serous fluid that, instead of forming a cheese-like mass, they are converted into an emulsion, which can be easily absorbed or expectorated. In this way there is no chance of caseation occurring, and, as a consequence, such persons do not suffer from pulmonary phthisis.

The smaller bronchi invariably show the appearances formerly described as indicative of acute catarrh of their mucous membrane, and the lobular septa in connection with these (Fig. 48 c, c, c), exhibit great cellular infiltration, the result of germination of their connective tissue corpuscles.

CATARRHAL PNEUMONIA: SECOND STAGE.

The second stage is that in which the catarrhal products in the air-vesicles caseate. The disease *always* passes through the first stage either acutely or sub-acutely. In children, after measles or whooping-cough, and sometimes idiopathically in adults, the first stage is extremely acute, but in other instances, more particularly in the adult, it is ushered in as a sub-acute disease, and bears this character throughout. The groups of air-vesicles are gradually filled with the catarrhal alveolar products, and the symptoms are correspondingly asthenic in character. The first stage ends with the accumulation and distension of the air-vesicles with catarrhal cells, and is followed, in the course of a few weeks, or in a shorter time, by the second stage, or that of caseation.

It has been remarked (p. 136) that the patient generally traces the commencement of his ailment to what he describes as "catching a cold," corresponding, as we have seen, to the first stage of the disease. If further interrogated as to the course which the ailment followed, some such expression as "the cold sank down upon my chest," will be employed to indicate that the first stage was not recovered from. This feeling of "sinking down upon the chest" corresponds to what I designate the second stage of the disease, or that in which the accumulated alveolar contents become caseous. Adults usually live on to the third or last stage of the disease, namely, that of phthisis, but they occasionally die before that is reached. Children, however, frequently die before any excavation or phthisis ensues. The naked-eye appearance of the lung in the second stage is the following :—

There is generally old fibrous adhesion of the two pleural surfaces, sufficient to obliterate the pleural cavity either partially or completely. The apex is the part at which this adhesion is usually most evident. Fibrinous effusion may be present on the non-adherent parts of the pleural surface. Irregularly rounded projections of the lung-tissue are noticed on the surface, and the organ when removed does not collapse. It is much increased in weight, and, on account of its remaining distended after removal, appears to be, and actually is, increased in bulk. When grasped in the hand, rounded nodules, corresponding to the above-mentioned projections, of hard consistence, and sharply-defined from the surrounding vesicular tissue, can be felt within it. When the organ is laid open, the most evident abnormal feature is the presence of these nodules. In the first stage of the disease the pneumonic portions were distinguished as *patches* of slightly infiltrated lung-tissue having a greyish-yellow colour, and from which catarrhal fluid could be squeezed out. Now, however, the patches have lost their indefinitely indurated character, and form hard *nodules* with a sharply-defined border. These nodules vary in size from a millet-seed up to that of a walnut; they are rounded in shape, and their border is somewhat irregular. They are dry on section, and have a cream-yellow colour. They appear to be little masses of tissue which have undergone caseous transformation. The smaller tend to run together to form larger nodules, and, occasionally, a great portion of a lobe may become continuously infiltrated by the further confluence of these.

The nodules are usually most numerous at the apex of the organ, but not by any means always so. They are relatively more abundant near the pleura than at the centre of the lung, and they spread out in a racemose manner, like a bunch of grapes. In the centre of such a group an occluded bronchus is sometimes seen, giving to it a still further resemblance to a bunch of grapes with the stalk in the centre. The intervening lung-tissue is vesicular and moderately congested. It is occasionally, but not usually, slightly œdematous; the prevailing character of the whole lung is that of great dryness. The nodules look like little tumours lying in a comparatively healthy lung

The bronchi are generally in a state of catarrh. Their

SECOND STAGE.

mucous membrane is red, and from their openings muco-purulent discharge can be squeezed out. The bronchial glands are almost invariably enlarged, and either show some grey markings in their interior, or they are yellow and caseous.

These yellow and dry nodules were called "tubercles" by Laennec, and were looked upon by him as specific growths, not merely as inflammatory products. He says (*Diseases of the Chest*, p. 253), "The matter of tubercles may be developed in the lungs or other organs under two principal forms, that of insulated bodies and infiltration." Nothing could have been more unfortunate than this designation of these bodies, for it is through this means that so much confusion has taken place in understanding the true nature of catarrhal pneumonia. They are even yet constantly described as "tubercle" nodules, "tubercular," or "tuberculoid" masses!

Let us examine these nodules a little more closely. A representation is given of one of them magnified fifty diameters, in Fig. 49, and a portion of a similar nodule, more highly magnified, in Fig. 50. When the low-power drawing is examined, it can be noticed that the nodule is made up of *a group of air-vesicles* distended with solid constituents. The outlines of the alveolar walls are still apparent, but the part is rendered totally non-vesicular by the presence within the air-sacs of solid material. This appears to be amorphous, granular, and peculiarly dusky or cloudy, and, when examined with transmitted light, has a brownish or dull grey colour. The reason why the infiltrated patch of lung-tissue assumes the characters of a nodule is, as seen in the drawing, that the effusion is confined to one isolated group of air-sacs, almost to the complete exclusion of those in the neighbourhood. On careful examination, it can be perceived that this group of distended air-sacs is occasionally attached to a terminal bronchus or to an infundibulum. In the centre of the nodule represented in Fig. 49 a distended infundibulum (*a*) is observed, and, had this been traced a little further upwards, it would in all probability have been found to be continuous with a small bronchus similarly occluded. It will be noticed that the solid effusion closely adheres to the alveolar walls, and being, as we have seen from the naked-eye examination, a hard substance, it would be a very difficult matter to dislodge it either by efforts of coughing or by other means. The group of air-vesicles is firmly packed with it, so that the

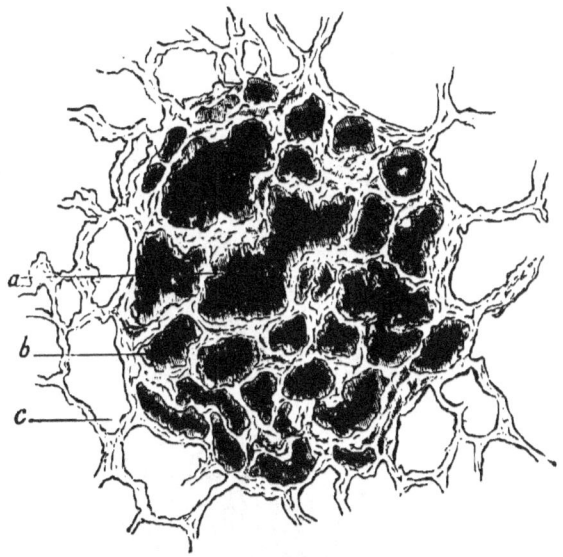

FIG. 49.—Catarrhal pneumonia, second stage, showing a caseous nodule magnified 50 diams. *a*, infundibulum filled with caseous material; *b*, air-vesicle distended with the same; *c*, neighbouring air-vesicle comparatively healthy.

alveolar walls and their solid contents may, practically speaking, be said to be continuous, giving the impression of a solid tumour when touched. It was this which, in the absence of accurate microscopic observation, misled Laennec and his followers, and caused them to look upon such nodules as neoplasms growing in the lung-tissue, and it is on this account that, even at the present day, they are so often looked upon as being tubercular.

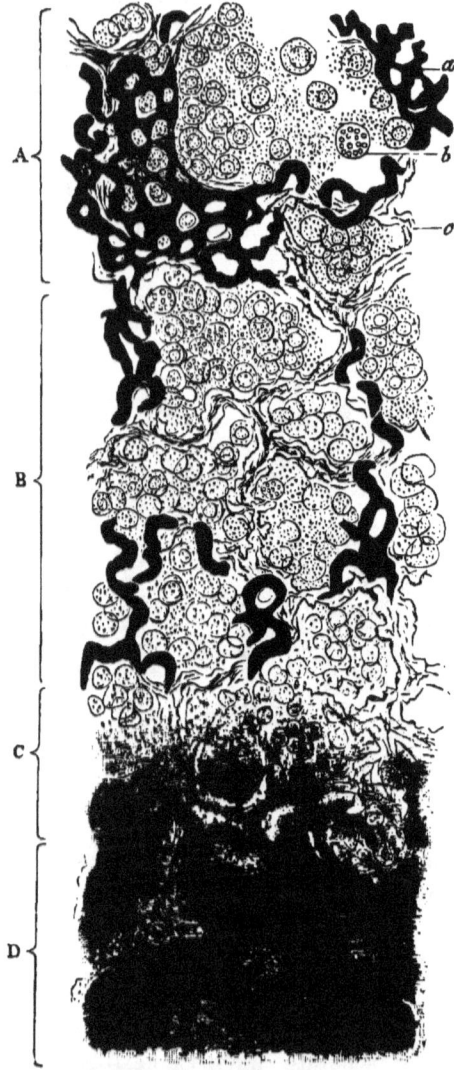

Fig. 50.—Catarrhal pneumonia, second stage, showing a nodule magnified 350 diams. A, B, C, and D, represent different areas in the nodule from the periphery towards the centre. *a*, injected capillaries of alveolar wall; *b*, catarrhal cells in alveolar cavities. *c*, an alveolar wall.

The surrounding air-vesicles are often in a state of acute catarrh. In Fig. 49 it will, however, be noticed that there is not any evidence of this, but that the adjacent pulmonary tissue appears to be comparatively healthy. This is frequently the case, the implication or not of the surrounding lung-substance depending on whether an intercurrent attack of acute catarrh has been present at the time of death. If this should have been so, then the surrounding air-vesicles show the same appearances as those figured in the different drawings representing the acute stage.

A more highly magnified view (350 diams.), not of the same but of a similar nodule, is given in Fig. 50. A segment of it is supposed to be cut out, the periphery towards A and the centre towards D. The drawing was taken from an injected preparation, and different areas, A, B, C, and D, are marked off, in order to indicate the progressive changes from the periphery towards the centre of the nodule. Suppose then that the area marked off at A, which of course would correspond with the periphery of the nodule, be first examined. It will be observed that in it the outlines of the air-vesicles (c) are still quite distinct. The capillary blood-vessels of the alveolar walls (a) have evidently been pervious, because the injection has run with ease through them. The cavities of the air-vesicles contain, but are not in this area distended with, catarrhal cells, similar to those previously described under acute catarrhal pneumonia. Some of these possess large nuclei, while others have been deprived of them, and are more or less fatty. Around the catarrhal cells there is a deposit of precipitated mucus, in which they are suspended.

As we proceed to the area comprised within the bracket at B, and which corresponds to a part nearer the centre of the nodule, these catarrhal cells become much more numerous, and now, instead of lying loosely in the alveolar cavities, they are closely packed together and distend them. Each group of catarrhal cells forms a little mass, the individual cells of which are closely united by mucus as before.

Many of the catarrhal cells are, however, undergoing disintegration, so that fatty and albuminous particles are set free, and these, mixing with the mucus already present, give it a highly granular appearance. It will be observed—and this is a most significant point, which can be verified either in an artificial injection, or in a natural injection with blood-corpuscles—that the pervious blood-vessels within this area are very much less numerous than in the area more removed (Fig. 50). All the nodules, in this stage of the disease, show this defect in the number of blood-vessels, as the centre of the nodule is approached.

When we pass to the area comprised within bracket C an additional change is observed. The catarrhal contents of the alveolar cavities have now lost their distinct form, and have become shrivelled, dusky, and granular, while the blood-vessels, which were still perceptible in area B although much diminished in number, have now entirely vanished. It is also evident that the alveolar walls are assuming the same dusky and granular appearance visible in their catarrhal contents. This dusky granularity is the microscopic evidence of commencing caseation, and shows that the part is dead or dying.

When the area D is reached, which corresponds with the centre of the nodule, the whole tissue is seen to have become completely caseous. The catarrhal cells are now no longer visible, but in their place there is an accumulation of dusky and cloudy caseous *débris*. The outlines of the alveolar walls have also undergone caseation, and have insensibly been fused, along with the alveolar contents, into an amorphous dusky and indiscriminately granular mass, in which the original lung-tissue can with difficulty be recognised.

Such then is the structure, from the periphery inwards, of one of these caseous nodules seen in the second stage of the disease. It is made up of a group of air-vesicles containing a few catarrhal cells at the periphery, filled to distension with them in the middle, and becoming caseous in the centre. The pervious blood-vessels progressively diminish from the periphery inwards, until, towards the centre of the nodule, they are absent. If we compare the appearances represented in this figure with those of Fig. 47, it will be perceived that the characteristic difference is that the catarrhal cells, in the second stage, are more closely packed in the air-vesicles, and that the centre of the infiltrated portion of lung-tissue has become the seat of caseous necrosis. Further, that while the vessels of the centre of the pneumonic patch can be injected in the first stage, they are impervious in the second. These changes are sufficient to cause the soft "frog's-spawn-like" pneumonic patch of the first stage to assume the consistence of a hard, yellow nodule in the second. The part, as it becomes caseous, loses some of its fluid, and at the same time acquires greater bulk, from the accumulation of the catarrhal cells; and hence the previously soft consistence which it had changes to that of a cheese-like hardness.

There is of course nothing special in the caseous degeneration of these pneumonic products, and of the lung-tissue in which they are contained. It is identical with what is so frequently seen in other parts of the body under similar circumstances. The process of caseation in catarrhal pneumonia is very much like that of the conversion of milk into cheese. The solids of the pneumonic effusion become relatively abundant; the fluid either drains off, or the solids increase in greater proportion than the fluid. These solids are composed of oily and albuminous constituents, which first undergo partial separation, and then become fused together into a structureless, cheese-like mass.

The condition under which caseation is liable to occur in any part of the body is *where the blood-supply is gradually cut off from the part.* If a terminal artery going to a part is *gradually* occluded, say by the process of thickening of the tunica intima, known as "arteriitis obliterans," it caseates, provided of course that anastomotic branches do not restore the circulation. Direct pressure, where the blood-supply is maintained, *does not* induce caseation, but gives rise to atrophy of the tissue elements, that is to say, to a mere separation and absorption of their particles.

SECOND STAGE.

Why is it, therefore, that caseation should be so liable to occur in catarrhal pneumonia? The reason becomes apparent on looking at an injected preparation (Fig. 50). We have seen that in the first stage the catarrhal products accumulate in the air-vesicles, and, if not removed, distend these, so as to press deleteriously on the alveolar walls. The area B (Fig. 50) shows this to be the case. The accumulation of the catarrhal products occurs gradually, and the pressure exerted upon the alveolar walls by them is consequently slowly applied. The circulation within the alveolar capillaries is thus progressively arrested so that they can be only partially injected (Fig. 50, B). As a result of this gradual deprivation of nutritive fluid the parts undergo a slow form of necrosis known as caseation, affecting those tissues first which have least vitality, namely, the catarrhal products, but finally also involving the alveolar walls. The fluid drains off from the part through the absorbents and other channels, and, as little fresh fluid is being carried into it as blood, it naturally becomes drier than formerly, and hence is not so liable to undergo rapid decomposition. If the blood-supply were *suddenly* cut off from a large area, and before this natural system of drainage had taken place, the chemical changes set up would be different, so that instead of the part caseating, it would undergo the moist form of disintegration known as gangrene. In the making of cheese the curdled milk is compressed, so as to drive out the milk serum. The solids of the milk which remain will then keep free from further chemical change for a very considerable period of time. The milk-curd, however, if allowed to retain its serum, would undergo almost immediate chemical decomposition. The pressing out of the fluid of the milk has the power of arresting chemical change, and of determining a different kind of decomposition when that does occur. This is popularly known as the "ripening of the cheese," an event which may be avoided for many months. These two processes are quite analogous to caseation and gangrene in the human subject. In the one, the serum is strained off from the oily and albuminous constituents of the tissues, and a drying of these consequently takes place. Under such circumstances, the caseous product, as it is called, may lie in a part of the body unaltered for many months, or even years. In a gangrenous tissue, however, much more active chemical decomposition is set up, on account of the moisture which it contains, and the process is accompanied by putrefactive fermentation.

Patches of lung-tissue affected with catarrhal pneumonia will be specially liable to become transformed into caseous matter, from the fact that the catarrhal solids *gradually* accumulate in the air-vesicles. Were the air-vesicles suddenly distended with them, there would be a greater likelihood of their becoming destroyed by a moist fatty metamorphosis. There is another reason, however, usually overlooked, why caseation should be of such frequent occurrence in the lung when affected with catarrhal pneumonia. In a person suffering from the acute or sub-acute attack, the irritability of the bronchial tubes is greatly increased, inducing, as it does, frequent attacks of coughing. In the act of coughing a full inspiration is taken, the glottis is closed, and the air within the chest is compressed by the muscles of expiration. In this way enormous pressure may be brought to bear upon any foreign contents of the alveolar cavities. If these should be partly composed of fluid, then this fluid will be driven through the alveolar walls, and will be removed by the neighbouring absorbents. In the first stage of the disease a considerable amount of fluid is contained within the catarrhal pneumonic secretion, which undoubtedly will be pressed out of the air-vesicles when the patient coughs, while the solids will be left within the air-vesicles in a more or less dry condition. This must therefore be a cause of the utmost importance in predisposing to caseation, and may perhaps account for the greater liability of this organ to undergo caseous metamorphosis than others which also suffer from catarrhal affections. In respect of this positive pressure exerted by the forced expiratory muscles in coughing, the lung may very appropriately be compared to the cheese-press used for the purpose of squeezing out the serum from the solid parts of the milk.

Such considerations make it evident that the more asthenic the course of the first stage is, the more liability will there be to the gradual accumulation of catarrhal products, and to their drying and caseating. If we could artificially induce an œdema of the lung where there is a suspicion of incipient caseation, little danger of this taking place would be incurred. The statement which I have before made, to the effect that caseous catarrhal pneumonia does not supervene in patients previously the subjects of mitral disease, bears this out, and the explanation undoubtedly is that the epithelial cells of the alveoli—although constantly being shed from the alveolar walls in such cases—mix with so much œdematous fluid that they cannot undergo the drying process necessary to cause them to caseate. The dry crackle of a caseating catarrhal pneumonia, as compared with the moist râles of an œdematous lung, sufficiently indicates the difference in the two kinds of alveolar contents. In the one case an inspissated viscid fluid is the cause of the sound, in the other it is due to a fluid containing much more water.

I have before indicated, and it must be a fact familiar to all, that caseation in catarrhal pneumonia generally commences at the apex. It is an equally important fact, however, and one not so generally recognised, that the pneumonic patches in the acute or first stage are not more abundant at the apex than at other parts, but are equally distributed throughout the lung, if anything, more numerously at the middle and base. It is those, however, that are situated towards the apex which are most liable to caseate, and to pass into a chronic state. Those at the base generally undergo a moist fatty resolution, and are removed in the first stage. How is this to be explained? The reasons are probably two in number. The first and most important is, that the apex is the driest part of the lung, while the base, from gravitation of fluids, is always, in an acute catarrhal pneumonic attack, more or less moist. As the catarrhal mucus is secreted at the apex, its fluid will tend to drain down to the lowest part of the lung, leaving the solids entangled in the air-vesicles. The catarrhal products at the base will therefore be more likely to undergo a moist fatty degeneration, while those at the apex will tend to undergo greater inspissation, and will thus be more liable to caseate. The other reason is, that the apex of the organ expands least. There is less motion at this part, and hence the catarrhal products have a greater tendency to be arrested in it.

The caseous nodules are more numerous at the periphery than at the centre of the lung, and this is accounted for by the inspiratory efforts drawing the catarrhal fluid outwards. The "bunch of grapes" arrangement, which the nodules frequently have, is due to the distribution of a small terminal bronchus connected with several different groups of air-vesicles. Each little area of lung-tissue, when distended with caseous products, figuratively represents a grape, while the bronchus with which its air-sacs communicate may be looked upon as the stalk.

When the pneumonic nodule has become completely caseous, it lies, as before mentioned, like a foreign body in the organ, to all intents and purposes a dead mass of animal tissue. Several changes may now ensue in the lung-tissue around it. The commonest is, that by repeated small intercurrent attacks of catarrhal pneumonia, in the vesicular lung-tissue between the caseous nodules, the neighbouring air-vesicles also become infiltrated with catarrhal products, so that the one nodule fuses with that next to it, and a larger mass thus results. In the course of time an entire lobe may become continuously involved in this way with caseous catarrhal deposit. It was to such that the name "infiltrated" or "crude" tubercle was given by Laennec, on the understanding that these yellow caseous masses were tubercular in their nature.

Another change noticed in the neighbourhood of these caseous pneumonic masses is, that the surrounding alveolar walls become thickened by formation of fibrous tissue, so that a spurious capsule is in this way formed. Such masses may be said to be encysted, and when this occurs they appear to be able to resist further chemical decomposition for a long time.

A third complication noticed around them is the effusion into the surrounding air-vesicles of a *croupous* exudation. This is to be expected occasionally, considering the principles on which such croupous exudations are formed. A great number of the blood-vessels in the lung are obliterated, from their being implicated in the caseous masses, and, of course, extra strain is put upon the remaining vessels which are pervious. In such cases a slight increase of the pumping action of the right ventricle will be almost certain to force into the pervious air-vesicles a fluid containing a high percentage of albumins, that is to say, a fluid rich in fibrin-forming materials. The only reason why croupous exudations in such lungs are not more abundant than they are is, that the obliteration of the vessels of the caseous nodules takes place gradually, so that the circulatory apparatus has time to accommodate itself to the new conditions, while the circulatory powers of persons affected with this disease are also unusually feeble. The extent of the croupous exudation in such cases varies, being, sometimes, general throughout a lobe; at other times, localised to a particular group of air-vesicles. When of a general character, it is of course a serious complication.

Pleurisy, more or less general, is almost invariably present in this stage. I have previously remarked that, in the first stage, pleurisy is not one of the usual features of the disease, in this respect forming a marked contrast with croupous pneumonia, in which it is seldom absent. In the second and third stages of catarrhal pneumonia, however, the pleuræ are always partially or completely adherent. The bond of union is either fibrous or fibrinous—not unfrequently both kinds of union are present, the fibrous being the remains of a former pleuritic attack, the fibrinous being the evidence of one more recent. The appearance presented on microscopic section of the pleura and neighbouring portion of lung in such a case is very instructive. The caseous and obliterated air-vesicles are seen scattered throughout the lung, with intervening portions of lung-tissue either comparatively healthy or containing some exudation. For about one-eighth of an inch of the lung-tissue adjacent to the pleura the whole of the alveolar capillaries are distended with blood-corpuscles. Branches come off from these, which run into the deep layer of the pleura and anastomose with its vessels proper. Both sets of vessels are similarly distended with blood. They further communicate with the vessels of the superficial layer of the pleura by a free anastomosis. The latter, however, instead of ramifying, as they usually do, in a horizontal manner, within the fibrous tissue of the membrane, are thrown out in loops, exactly as on a granulating surface, while round them there are cells of a granulation character. In many places the vessels have forced their way through the pleural endothelium, and have carried with them the surrounding granulation cells. It is by the coming in contact of two such parts from adjacent sides of the pleural cavity that adhesions take place. On the surface of the membrane, and in the fibrous meshes of the pleura, much fibrinous exudation, with leucocytes in it, is visible, while the granulation vessels also carry with them numbers of larger cells, derived either from the pleural connective tissue or from the pleural endothelium. By the organisation of these, the permanent fibrous union is accomplished. The whole appearances are indicative of undue pressure applied to the pleural and pulmonary capillaries. The pleural surface becomes, practically speaking, a granulating surface, the granulation loops projecting into the effused pleural fluid. When this fluid is absorbed, the two pleural surfaces come in contact, and permanent adhesion takes place. The cause of the undue vascular pressure which occasions the pleurisy is the same as that which sometimes forces a croupous exudation into the lung, namely, that part of the capillary surface of the lung being destroyed by the caseous degeneration, too great a strain is brought to bear upon those vessels which are still pervious. This reacts upon the pleural vessels, causing increased exudation and the construction of a false or fibrinous membrane, followed by adhesion. As the lung becomes more and more implicated in the disease, that is to say, as the third stage is reached, the pleura is still further involved, until, as usually happens, the pleural cavity is totally obliterated. I cannot look upon the pleurisy, which is invariably associated with catarrhal pneumonia in its second and third stages, otherwise than as a manifestation of the effects of diminishing the circulatory capacity of the lung.

Hæmoptysis is met with in this stage, and the cause of it undoubtedly is a further increase of that blood-pressure which induces the occasional croupous exudation. It is due to large tracts of lung-tissue being practically useless for purposes of blood-circulation, so that, on any increased excitement of the heart, the tension of the blood in those vessels which are still open becomes sufficiently raised, not only to cause the transudation of fibrin-forming solids, but actually to give rise to rupture of their over-strained coats. Any unusual exertion, such as running or other muscular effort, by which the pulmonary circulation has increased demands on its capabilities, will be sufficient to determine this. Were all the circulating channels quite free, no ill-effects would be experienced, seeing that the venous outlets are always more than sufficient to carry off any amount of blood that may be pumped into the organ by the arterial inlets. But when, as has been described, the caseous tracts destroy a large part of the capillary channels, it is only natural that if an increased quantity of blood be driven into the organ. the strain thrown upon the still pervious vessels will be sufficient to produce either exudation of fibrin-forming solids, or, if carried further, actual rupture and hæmoptysis. The same principle is at work here as in other similar circumstances. Wherever the venous outlets of the circulation in a part are insufficiently large to accommodate any sudden increase in the quantity of blood poured into them, one of three things will follow,

M

according to the amount of strain they suffer—either the part will become œdematous, or a croupous exudation will be poured out, or actual rupture of the engorged vessels will follow.

Where the pneumonic masses are not very numerous they may become inert, by a process of calcification. The caseous matter seems to be gradually absorbed, and an impregnation of the part with calcareous salts follows. It is in nodules which are surrounded by a fibrous capsule that this is most usually seen.

If the patient survive long enough, however, the commonest sequence noticed in these caseous catarrhal nodules is softening, and it is this which constitutes what I call the third stage, or that of phthisis. Following the order of description of the different stages that I have pursued, the third stage ought to be next taken into consideration. As, however, it would be impossible for me to make this intelligible without first explaining what tubercle of the lung is, it will be necessary to postpone the investigation of the phthisical stage of catarrhal pneumonia till afterwards. I shall therefore, as a preliminary measure, crave the reader's attention for a short time to the subject of "Tubercle in the Human Lung."

Tubercle in the Human Lung.

Under whatever circumstances tubercle is found in any organ, in man or in the lower animals, one conditional law always holds good as to its development. It is invariably preceded by some source of infection, usually a caseous deposit, either in the tubercular organ or in some distant part which is in process of softening. Old as this statement is, it has occasionally, of late, been called in question, either from the caseous source of infection not having been detected, or what is quite as likely, from the original caseous deposit having been mistaken for the tubercle itself.

It is now many years since Villemin (*Comptes Rendus*, vol. lxi. p. 1012) showed that in rabbits an eruption of miliary nodules, in all respects similar in their general characters to so-called miliary tubercle in man, could be induced within a few weeks by the inoculation of cheesy deposits in different parts of the body. The cheesy material evidently contained something which, being absorbed, had the power of creating multiple small tumours in distant parts. There is every reason to believe that tubercle in man is induced in the same way. The softened débris of a caseous tissue, when absorbed by the blood-vessels or lymphatics, appears to have the power of producing multiple small tumours throughout the system. Some material, apparently a ferment, is elaborated in the process of caseation, which acts as a specially virulent irritant upon the tissue into which it is carried, exciting that tissue to abnormally great activity of growth, and producing the little neoplasm which we name tubercle. There does not seem to be any preference in the seat of the primary caseous deposit, any tissue *under certain conditions,* so far as my experience goes, being capable of inducing the formation of tubercle when it caseates.

Both the lymphatic channels and the blood-vessels appear to be capable of absorbing the softened caseous products, and of disseminating them throughout an organ or throughout the system generally. There is every reason to believe that it is originally in connection with the tissues of these two sets of vessels that the tubercles arise, and each seems equally capable of being influenced by the caseous irritant. If the tubercles are locally produced within the same organ as the primary source of infection, the lymphatic channels are usually the seat of their development. But if the tubercle nodules are not confined to the same organ as the infecting source, if they are widely spread abroad throughout the system, the blood-vessels are to be looked upon as the means of conveying the poison.

The caseous source of infection may be at some distance in a neighbouring organ, and the tubercles alone may be present in the lung, or both the caseous source of infection and the tubercles may cotemporaneously exist within it, the one being the cause of the other. To the former, that is to say where the caseous source of infection is in some distant part or organ, and where the tubercles alone are present in the lung, I shall give the designation of "primary tubercle." To the latter, where the tubercle nodules and the caseous source of infection are together present in the lung, I shall give the name of "secondary tubercle." In relation to the lung, the tubercle in the former is the primary, and, in fact, usually the only disease. In the latter the caseation, from whatever cause arising, is the primary disease, and the tubercle is consequently secondary.

Primary Tubercle of Lung.

It is generally presumed that tubercle of the lung is commoner in children and in youth than in adults. With this statement I cannot agree *in toto*, but would qualify it by saying that *primary* tubercle of the lung is more common in childhood and in adolescence than after the age of twenty-one years, but that *secondary* tubercle is certainly more frequently seen in persons above than below the age of twenty-one. Of the two, secondary tubercle is the commoner. It is therefore incorrect to say that tubercle is *par excellence* a disease of childhood and youth more than of middle life. The remark is true of primary tubercle of the lung, not of secondary.

The reason why primary tubercle is seen in the lungs of children and adolescents rather than in persons older, apparently is, that glandular enlargements with caseous degeneration are more common in the former than in the latter. That strumous subjects are much more liable to glandular enlargements before the age of twenty-one than after it, is a fact familiar to every one. It is these softening strumous affections of glands which are the great source of primary tubercular formations in the lung and other organs, and hence it is that this variety of the disease is so frequent in youth and in childhood.

It is of the utmost importance before commencing to reason on a supposed instance of tubercle in any organ, and more especially of tubercle in the lung, that there should not be any doubt as to the tubercular nature of the bodies under consideration. So close is the superficial resemblance between tubercle and the tubercle-like affections of the lung, that unusual care is necessary in their investigation. The mere naked-eye examination of a supposed tubercular structure in the lung is not to be looked upon as at all conclusive. I would never pronounce a body to be tubercle in the lung without confirming the macroscopic by microscopic examination. With the microscope the first glance is sufficient to enable the practised eye to detect the true nature of the nodule under consideration. Of all the neoplasms tubercle is perhaps the best defined, so that there ought not to be any ambiguity in the meaning of the term.

Primary tubercle of the lung is usually called "miliary tubercle," on account of the supposed resemblance of the tubercle nodules in size and shape to millet-seeds. The word "miliary" is employed in antithesis to a supposed variety of the disease in which the nodules have not the miliary character. There is not in reality, however, any difference in the shape or general character of the tubercle nodule in any form of the disease; and hence it would save much misunderstanding if this relic of a bygone system of nomenclature were forgotten, and the term tubercle retained in the pure sense of indicating a tumour with certain specific and constant characters.

On the same grounds it might be asked, why retain the name "tubercle" at all? It certainly has not a very definite signification when taken in its literal sense, but having been used for so long it will be difficult to get rid of it. It is, moreover, after all not very misleading, for seeing that it was originally applied to designate a similarity in the nature of the nodules constantly found in different organs in cases of general tuberculosis, it has on that account a certain specific meaning. It is the inclusion of mere caseous deposits under this term that has caused so much confusion. So long as it is employed in reference to deposits which in their structure and mode of production are alike, it answers all practical purposes. The great danger may be that it is taken in too wide an acceptation, being applied to any tissue which has become caseous. The same confusion formerly existed in regard to the cancerous and sarcomatous tumours, which we now know to be entirely different in their structure and mode of origin. We are also in a position to assert that tubercle has a perfectly distinct and constant structure different from any other known neoplasm, and the name "tubercle" sufficiently expresses this. The sooner, however, such words as "miliary," "crude," "infiltrated," "tuberculoid," and a host of others are removed from the category of pathological terms, the sooner will there be a likelihood of gaining clear and distinct ideas on the subject of the disease itself. The abnormalities of any organ are comparatively easy to understand; their nomenclature is frequently incomprehensible.

In the following description of primary tubercle in the lung I shall take as my guide a typical instance of the disease, in which its commencement, course, and duration were accurately known. The subject of it was a woman aged twenty, who was delivered of a child thirty-three days before death. Previous to this she had from all accounts been in good health. On the seventh day after delivery she suffered from a rigor, followed by considerable fever, which continued from this time up till that of death. It was also evident from the symptoms that the patient was suffering from peritonitis. The only pulmonary symptoms were those of slight bronchitis. She rallied to a certain extent at one period of the disease, but about seven days before death became worse, and finally died with signs of cerebral meningitis.

The dates of the case are important as bearing upon the age of the tubercular deposits found in the lungs: her illness

TUBERCLE IN THE HUMAN LUNG.

dated from a week after delivery, and twenty-five days elapsed from this before death took place ; so that, even at the utmost, a month had been sufficient to induce all the morbid appearances to be described, and the tubercular growths in the lung could not have been more than from a fortnight to three weeks old.

After death there was found to be extensive peritonitis, of quite recent and also of somewhat older date. In many places, more especially behind the uterus, the peritonitic lymphy effusion had become *caseous*, and here and there this had undergone *softening*. Covering the peritoneum, more especially in regions adjacent to the softening caseous effusion, there were large numbers of grey tubercles in the peritoneum, running in lines along the course of the lymphatic vessels leading to the under surface of the diaphragm. Nearly all the organs showed tubercles of recent origin, and there was extensive cerebral tubercular meningitis. The lungs presented appearances typical of those usually found in cases of primary tubercle, and it is specially to these that I must now direct the reader's attention.

There was not any evidence of recent pleurisy. In looking over the post-mortem accounts of many similar cases of primary tubercle of the lung, I find that the records in regard to this particular vary. In certain instances there has been a little fibrinous deposit on the pleura, but, usually, it has either been absent, or has been present in comparatively small amount. A little roughening and dulness of the pleural lustre has been the utmost I have observed in any such case.

The pleura was beset with tubercle nodules, and when incised they could be seen to lie in its deep layer. Similar tubercle nodules were seen in immense abundance in both lungs, uniformly distributed throughout all the lobes. They had the same characters as those seen in the pleura. They were round and had a sharply-defined border, which was abruptly marked off from the surrounding pulmonary parenchyma. Their colour was grey, and they had a somewhat gelatinous aspect. They were about the size of a mustard seed. This is generally the case both in primary and in secondary tubercle of the lung. Bodies having an aspect similar to that of tubercle, but larger than a mustard-seed, usually prove to be groups of air-vesicles in a state of catarrhal pneumonia. All the nodules were, as near as possible, of the same size. They either ran in lines along the course of a small branch of the pulmonary artery, or they were aggregated in little clusters. The former was the commoner of the two arrangements. There was not any evidence of the nodules uniting to form larger nodules : for even although they might occasionally be seen in groups, yet the individual members of the group, after their border was once defined, never coalesced so as to constitute a single mass, and they never increased in size beyond the dimensions above stated.

The nodules, more especially in secondary tubercle of the lung, may in certain cases become united and surrounded by fibrous tissue, but even here they never fuse together, and if anything rather decrease than increase in size. The mere fact of certain tubercle-like nodules running together is strong evidence of their non-tubercular nature.

Some of the tubercles occasionally had a slight cream-yellow colour, but this was not their usual appearance, a grey connective-tissue-like aspect being that which was most general. There was a total absence of blood-coloration within them, and hence they stood out prominently from the highly vascular background on which they lay.

The intermediate lung-tissue was vesicular, and did not readily collapse when incised. As in acute catarrhal pneumonia, some parts of it were slightly emphysematous. It was much congested, and the blood had a bright scarlet colour. The cause of the congestion in such cases, as we shall see, is the delay in the transmission of the blood through the organ, caused by the presence of the tubercles. The bright colour of the blood is due to hyper-oxygenation from the same cause. It is generally supposed that if the blood is hindered in its course through the lung it becomes of a cyanotic tint. This is manifestly erroneous, if air is freely entering the lung, for it is evident that the longer the blood is delayed in its circulation the more oxygenated will it become.

In the case which has been recorded as illustrative of primary tubercle, the commencement of the disease evidently was peritonitis following delivery. Caseation ensued in the peritonitic effusion; this softened; the caseous débris was absorbed, and gave rise to the formation of tubercle in various organs. The tubercles in the peritoneum were evidently of local formation, and had their origin in the lymphatics. In the other organs, there is every reason to believe, as will be shown, that the blood-vessels were the means of transmission of the caseous products, and that the tubercle was formed within them.

The bronchial glands in primary tubercle are not always enlarged, while in secondary tubercle they invariably are much increased in size, and contain many tubercles. In primary tubercle of the lung the occurrence of enlargement of the bronchial glands, from the presence within them of tubercle, is to be looked upon more as an accident than as a necessary concomitant. The cause of their enlargement in secondary tubercle will be made evident when that form of the disease is considered. It may suffice at present to state, that the fact of the presence of tubercles in the lymphatic glands in secondary, and of their occasional absence in primary tubercle of the lung, is quite in accordance with the mode of propagation of the tubercle nodules in each.

I have given, in Fig. 51, a representation of a low-power (50 diams.) view of one of the tubercle nodules taken from the above related case of primary tubercle. It can be noticed, although the nodule is not more than from two to three weeks old, that it is, even at this early date, quite distinctly defined, and that its borders are sharply marked off from the surrounding air-vesicles. The whole character of the tumour is that of an interstitial growth which is pushing the pulmonary tissue aside, and flattening the air-vesicles adjacent to it. On one side is a portion of the wall of a branch of the

170 TUBERCLE IN THE HUMAN LUNG.

pulmonary artery (*a*), while small offshoots of the same (*d*) are seen in its neighbourhood. The nodule lies close to the larger branch of the vessel.

The tubercle, as will be seen, is, in this stage, in great part cellular, with a delicate fibrous stroma running through it, while in the centre caseation (*c*) is commencing. Within the nodule several cells of large size (*b*), which are easily recognised with this low magnifying power, are to be noticed.

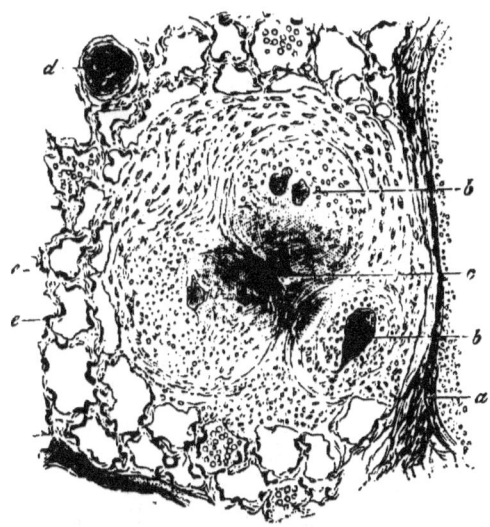

Fig. 51.—Primary tubercle of lung, two to three weeks old, × 50 diams. Source of infection was a caseous peritonitis. *a*, portion of wall of a branch of the pulmonary artery; *b*, giant cells with concentric arrangement of fibrous tissue; *c*, centre of tubercle beginning to caseate; *d*, small branches of pulmonary artery seen on transverse section; *e*, injected capillaries of the alveolar walls.

They are the giant-cells, which form so important a diagnostic feature of all tubercular growths. Even with this enlargement it is apparent that these giant-cells vary much in size and contour, and in the high-power drawings given of several of them (Figs. 53, 54, and 55) these characters are more evident. Around each giant-cell or group of giant-cells a delicate concentric arrangement of fibrous tissue is apparent, so as to give rise to the impression that each giant-cell forms a nucleus for a separate tubercle or giant-cell system (Fig. 51).

All the large tubercles in this lung were totally non-vascular. The organ was injected, but in no case was the injecting fluid found to penetrate into the tubercle growths. We shall afterwards find that, from the mode of development of primary tubercle of the lung, the destruction of the blood-vessels in the part is virtually necessitated. It is for this reason that caseous degeneration ensues so quickly in the centre of the growth. The tumour being non-vascular can only be supplied with nourishment from the vessels of the lung at its periphery. Whenever, therefore, it grows to such a size that the nutritive fluids so supplied cannot reach its innermost parts, caseous necrosis ensues, and the centre of the tubercle dies. The caseation always commences at the centre, and the cause of this is evident, the central portion is the furthest away from the nutritive supply.

By the time that the tubercle becomes an object visible to the naked eye, giant-cells can always be detected in it; and this holds good not alone for tubercle of the lung, for in tubercle of all other organs they are invariably present whenever the nodule becomes defined. It is evident from the account of the above case, and from many others which might be cited having a similar history, that well-developed tubercles may originate within a space of from two to three weeks. Tubercle was formerly considered to be a chronic disease, and the fact of giant-cells being developed in so short a time as three weeks has partly led M. Charcot to the conclusion that catarrhal pneumonia and tubercle of the lung are alike. I trust to show that this idea is totally erroneous and extremely misleading.

When we find that tubercle growths can be artificially developed within a few weeks in animals, as a result of injection of caseous material, there is not any difficulty in seeing how they may originate in a similarly short space of time within the human subject, where, practically speaking, the conditions essential for their production are the same. Secondary tubercle of the lung, no doubt, is occasionally a very chronic disease, but the primary form runs an acute course, the tubercle-nodule being well developed in from a fortnight to three weeks.

Looking at the tubercle represented in Fig. 51, and comparing it with a nodule of catarrhal pneumonia, such as that seen in Fig. 49, there cannot be the slightest difficulty in recognising the essential differences between the two. In the earlier stages of development of a primary tubercle-nodule there is a certain resemblance to catarrhal pneumonia, but, even then, there are inherent features which will be shown to exist in the tubercular growth, marking it as such from the commencement.

Before going further, it will be advantageous to examine in detail what the histological constituents of a tubercle are when it has reached its maximum stage of development. In Fig. 52 a representation of such a tubercle is given, magnified about 450 diams., taken from the lung of an adult. Throughout the tubercle several giant-cells ($a, a, a,$) can be noticed, their size entitling them to the designation given to them. Compared with the "lymphoid" cells (f) shown at different parts of the figure, they are seen to be from ten to thirty or forty times larger. They are sometimes situated in the centre of the tubercle-nodule, at other times they are placed laterally.

These giant-cells, occurring as they do with the utmost regularity in all tubercular growths, are among the most remarkable of histological structures. Their form is varied, and those seen in Figs. 53, 54, and 55 will give the reader an idea of the shapes which are commonest.

While the cell is young, it seems to consist simply of a large mass of very granular protoplasm, sometimes with many nuclei in it, but at other times devoid of these. As it grows older, however, the periphery becomes organised, forming a fibrous mantle-like sheath, in which great numbers of nuclei of an oval or round shape can occasionally be

TUBERCLE IN THE HUMAN LUNG. 173

perceived (Fig. 54, *b*). These nuclei are frequently arranged in a crescentic manner at one end of the cell (Fig. 53, *c*); at other times they form an almost complete covering for it. The cell, when it has reached the size of those depicted in the figures, therefore consists of two distinct parts. The

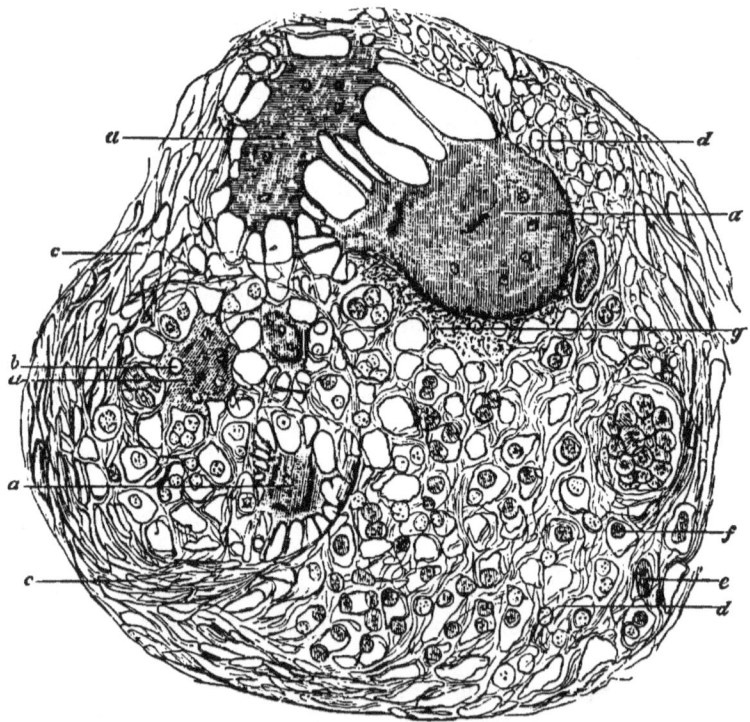

FIG. 52.—Fully developed tubercle of lung, ×450 diams. *a, a, a*, giant-cells; *b*, vacuole in one of these; *c*, peripheral capsule of fibrous tissue; *d*, reticulum of the tubercle; *e*, large endothelial-like cells lying on the reticulum and within its meshes; *f*, smaller "lymphoid" cells occupying the same situations; *g*, peripheral fibrous-looking periplast of the giant-cell.

central portion is granular and protoplasmic, the peripheral is somewhat fibrous in character, and contains many nuclei. A vacuole is sometimes perceptible in the central protoplasmic part, but otherwise it seems to be uniformly granular. I have

174   TUBERCLE IN THE HUMAN LUNG.

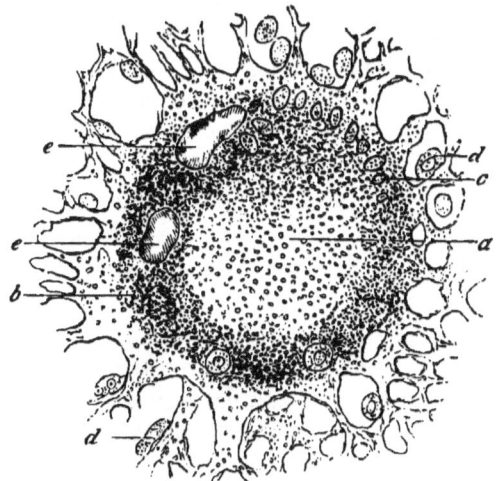

FIG. 53.—Giant-cell from centre of tubercle of lung, ×430 diams. *a*, granular protoplasmic centre ; *b*, peripheral more formed periplast ; *c*, crescent of nuclei ; *d*, endothelial-like cells ; *e*, two vacuoles within the giant-cell.

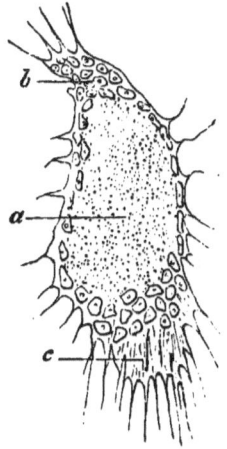

FIG. 54.—Large oval giant-cell from tubercle of lung, ×300 diams. *a*, granular centre ; *b*, periplast, with nuclei, forming a mantle-like sheath ; *c*, processes of periplast.

searched for a protoplasmic plexus within this portion of the cell, but as yet have been signally unsuccessful.

In giant-cells found within well-nourished tubercles, in which of course their ultimate development is to be looked for, a still further process of differentiation can be noticed in the peripheral part, in that it spreads out as a delicate film (Fig. 53), which soon begins to show evidence of differentiation into fibres. The fibrillation of this peripheral film is, at first, very delicate, but soon becomes easily visible. The nuclei, which formerly were arranged round the central protoplasmic portion of the cell in the form of a mantle, are now spread out on the surface of this fibrous periphery, very much like the endothelial nuclei on a flat membrane. In all respects it seems as if the giant-cell of tubercle in old tubercles

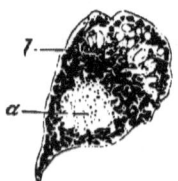

FIG. 55.—A giant-cell from tubercle of lung, with inhaled particles of carbon in its interior, ×300 diams. *a*, central granular part; *b*, peripheral pigmented portion.

possessed the power of elaborating from its periphery a substance which ultimately becomes transformed into white fibrous tissue.

At a still later period the fibrous appearance just described becomes more defined, and then a separation into distinct bundles, or, as they are called, the *processes*, of the giant-cell ensues, which, as they stretch outwards, divide and subdivide apparently by splitting and contracting until a reticular tissue is produced (Fig. 52, *d*). The original nuclei of the peripheral portion of the cell now enlarge and lie flatly on the reticulum (Fig. 52, *e*), or they are contained within its meshes.

All this fibrillation and fibrous organisation of the periphery of the giant-cell has been accomplished at the expense of the central protoplasmic area. This progressively diminishes in size as the peripheral portion becomes more differentiated, so that ultimately the giant-cell comes to have the appearance represented in Fig, 56, where the central protoplasmic part (*a*) is reduced to a minimum, while the peripheral part has been converted into a well-defined flat fibrous membrane (*b*), or into a reticular fibrous tissue (*c*).

In the course of time the whole of the central part of the giant-cell is destroyed, being converted into a reticular fibrous tissue by a continuation of the process above described, while

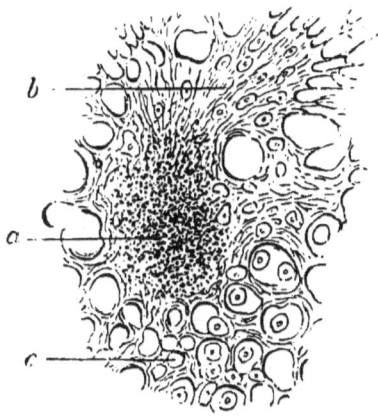

FIG. 56.—Remains of a giant-cell in process of fibrous transformation, ×400 diams. *a*, some of the central protoplasmic part still remaining; *b*, the peripheral portion, or periplast, which has now become developed into reticular fibrous tissue—nuclei are seen lying upon it; *c*, the reticulum constituted by the periplast.

the original nuclei of the peripheral mantle-like sheath remain upon the fibrous walls of the meshes of the reticulum as their connective-tissue nuclei. The further effects of this fibrous transformation of the giant-cells on the tubercle-nodule will be traced when the degenerations to which tubercle is liable are described. In the meantime this preliminary statement as to the evolution of reticular fibrous tissue from giant-cells will serve to render the structure of tubercle intelligible, and will show what the genetical significance of the central protoplasmic part of the giant-cell is in relation to the more highly organised periphery. The one represents an embryonic formative tissue, the other stands in the relation of a connective-tissue periplast. The mode of development of the one from the other corresponds to that pursued in the construction of all fibrous tissues.

A tubercle, therefore, when full grown, consists of a small, rounded tumour, made up of one or more giant-cells, from which multiple fibrous processes radiate, and constitute a reticular fibrous tissue. The walls of the reticulum are distinctly fibrous in character, and, when carefully examined, are seen to be flat membranes, on which nuclei lie, very much as on any fibrous connective tissue. These nuclei sometimes become detached, and are thrown off into the mesh of fibrous tissue, thus accounting for the lymphoid and epitheloid cells, which have been so often described as elements of the tubercular neoplasm (Fig. 52, *e* and *f*). The variety in size and shape of these lymphoid and epitheloid cells of tubercle is so varied as to preclude the idea of their all being blood-leucocytes, as has sometimes been supposed. The largest of them are known as small giant-cells (*sic*), on account of their dimensions, while the smallest are about the size of a blood-leucocyte; all possible gradations in size between these two are perceptible. They probably have the power of wandering through a certain area, but their capabilities in this respect must be limited, on account of the close-meshed reticulum in which they are contained.

I would consequently look upon the greater number of the lymphoid and other small cells found within a tubercle as endothelial in their nature, and as in reality representing the connective-tissue corpuscles of the fibrous reticulum on which, or within which, they lie.

Towards the periphery of the tubercle the meshes of the reticulum become closely pressed together and flattened, so that its walls constitute a fibrous capsule, as it were, in which the giant-cells and other elements are inclosed (Fig. 52, *c*). This forms the boundary of the tubercle or "giant-cell system," as it may be called. A tubercle-nodule is made up of a single such "giant-cell system," or, what is quite as common, of two or three of these "systems" combined. When the latter is the case, the different "systems" are bound together by fibrous tissue in which there are usually many nuclei. The older the tubercle is the better developed does the fibrous capsule at the border become, and, as it grows thicker, we shall see that it encroaches upon the interior of the tubercle, and is ultimately accompanied by absorption of most of the cellular elements.

In primary tubercle of the lung, of the age I have supposed in the case previously related (two or three weeks), it is seldom that the giant-cell systems are so completely developed as in secondary tubercle of the lung, where the disease runs a chronic course, and where, in addition, there is not the same liability to rapid caseation. If caseation should occur while as yet the tubercle is in an embryonic state of development, its structure is of course destroyed, and its further progress is necessarily cut short. In primary tubercle of the lung we could not therefore expect that the "giant-cell systems" should be so characteristic as in the secondary form, but, in all cases, the giant-cells, at least, and the same tendency to the peculiar development above described, can be easily detected (Fig. 52). In certain nodules, moreover, where caseation has not occurred at an early period, the "giant-cell system," in all its points, is perfectly developed even within this short period.

In every tubercular organ, if sufficient time has elapsed, the structure just described as characteristic of tubercle of the lung is to be seen either completely or partially developed. We accordingly have, in the histological composition of tubercle, a sure method of diagnosing those tumours which are tubercular, and of rejecting those which, although they may resemble tubercle in their gross characters, nevertheless prove to be totally different in their nature when more minutely examined.

Not only has tubercle a common structure in all situations

in the human subject, but in the lower animals a structure identical with that which is called tubercle in man also prevails. In horned cattle, the disease is perfectly well known.[1] In the latter, it is so frequent as to form one of the scourges to be dreaded by cattle-breeders. It is known by various unscientific names in this country, but the name given to it in veterinary medicine is now usually "tubercle," showing that the close resemblance to the human disease has been recognised. In Germany it goes by the name of "Perlsucht" or "Franzosen-Krankheit." It was long thought to be something different from the human disease by continental pathologists, but recent observations have gone to establish their identity.

I have made very careful examination of the organs in several instances of this disease, and feel the utmost confidence in stating that in all respects the two diseases— human tubercle and the "Perlsucht" of oxen—are identical. They are identical in structure, and, in their modes of development, follow the same laws. Another point which still further confirms their identity, is that the true "Perlsucht" tumours are preceded by a softening caseous source of infection. The same misunderstanding exists in regard to this, however, which for long obscured, and, to some extent, still complicates the subject of tubercle in man—namely, that the caseous source of infection is mistaken for the tubercle itself. The infecting source may be situated in different parts of the body, in any tissue. It sometimes presents itself as a primary affection of bone, and in pleurisy a fertile source of caseation and tubercle propagation is to be found. The large caseous deposits seen in the pleura, or in the bronchial glands of oxen after pleurisy, are by many veterinary surgeons mistaken for tubercle, just as they were by practitioners of human medicine in former times. They, however, do

[1] I am informed by Mr. Williams, Principal of the new Veterinary College, Edinburgh, that caseation is more liable to occur in the organs and tissues of oxen than in any other of the domestic animals. This probably accounts for tubercle being such a common disease in them.

not represent anything more than tissues which have undergone a particular form of necrosis, and they are totally different from the tubercles which may arise from them.

In rabbits and guinea-pigs a similar disease occurs, and can be artificially produced by the injection of caseous material from the human subject. In the former, tubercle arising from a caseous catarrhal pneumonia is a common cause of emaciation and of death. The tubercle-nodules here virtually have the same structure and appearance as in the human subject.

We are, therefore, justified in saying that tubercle is a tumour having a definite structure in different genera of animals, which is easily recognisable by microscopic examination. Where this structure is not present in a suspected tumour we have no right to call it by the name of "tubercle."

Having described in detail the histology of tubercle of the lung, it will now be advantageous to examine the method by which the primary form of it is developed. After investigating many instances of primary tubercle of the lung, I find that in all of them the first thing noticed is a little cellular projection on one side of an alveolus. From the fact of tubercle arising in connection with the absorption of caseous *débris*, it might be supposed that embola should be detected in the vessels as the primary lesion. I have never been able, even after the most diligent search, to persuade myself that such is the case, probably from the absorbed caseous materials being small in size and amorphous in shape. Where three alveoli lie adjacent to each other, the embryo tubercle may project into the whole of them simultaneously. This projecting cellular mass is caused by an interstitial thickening of the alveolar wall, so that it can be easily understood how it may protrude into several alveoli simultaneously. It is not the result of a primary germination of the alveolar epithelium.

At first, the elevation is only faintly perceptible, but soon the alveolar thickening increases to such an extent that the previously sessile tumour comes to have a definite shape, and

TUBERCLE IN THE HUMAN LUNG.

is somewhat pediculated. It grows into the alveolar cavity, and, as it does so, carries the alveolar capillaries with it. The alveolar epithelium can be seen, while the tumour is small, stretched over its surface. Fig. 57 shows a tubercle in this stage. On one side is represented a large branch of the pulmonary artery (f) injected with Prussian blue,

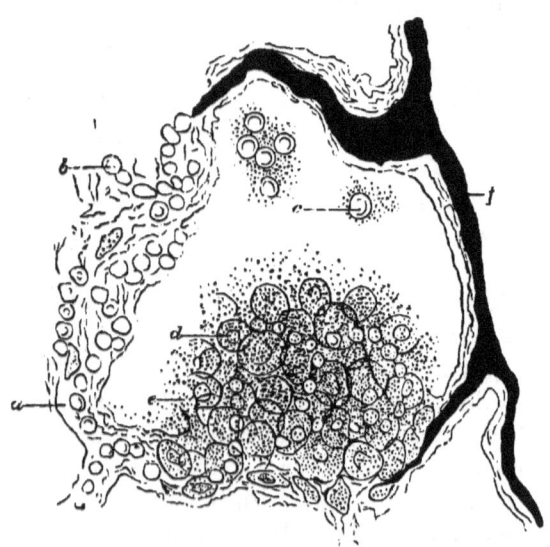

FIG. 57.—Primary tubercle of the lung in a very early stage of development, × 400 diams. *a*, an alveolar wall; *b*, blood-corpuscles in capillaries of the same; *c*, blood-corpuscles extravasated into the alveolar cavities; *d*, alveolar capillaries filled with blood-corpuscles carried forwards by the tubercle, which is growing into the alveolar cavity; *e*, large endothelial-like cells, of which the tubercle in this stage is mainly composed; *f*, branch of the pulmonary artery injected, the injection terminating abruptly in the alveolar branches.

while the alveolar walls are shown at *a*. A group of germinating cells (*e*) is seen firmly attached to, or rather incorporated with, the interstitial tissue of the air-vesicle, and projecting into its cavity. Capillary blood-vessels (*d*) filled with blood-corpuscles, have been drawn into it, and in this stage are still distinctly visible. They are all much engorged, and occasionally minute extravasations are visible, the blood-corpuscles being thrown into the alveolar cavity (c).

The cells of which the tumour at this time consists are large spherical or flat bodies with a well-defined nucleus (e). They have a granular appearance, and seem to be actively germinating. They are exactly like the connective-tissue nuclei seen on the surface of œdematous and proliferating fibrous tissue, and, so far as I have been able to trace their origin, they seem to be formed either from the connective-tissue elements of the alveolar wall, or from the endothelium of certain of its capillaries. Both of these appear to be sometimes drawn upon as formative sources, and from the similarity in their nature there is no reason why they should not act as generating tissues. One thing is certain, both in primary and in secondary tubercle of the lung, namely, that the tumour originates from a connective-tissue structure, and bears this character throughout.

It can be easily seen in Fig. 57, both from the abrupt manner in which the injection ends in the artery (f) as well as from the engorged state of the capillaries beyond, that there has been considerable retardation in the flow of blood through this part of the lung. When we consider how the capillaries are pushed or dragged into the alveolar cavity by the cellular mass the cause of this becomes evident. The tubercle growth in this stage, as will be observed, somewhat resembles a granulation, and, of course, where the alveolar capillaries are drawn out into this convex shape, and where they are attenuated, the blood must have great difficulty in circulating through them. The bright scarlet colour and congested appearance of the lung in primary tubercular disease are thus accounted for. The blood is delayed in its circulation, and air freely entering the organ, it becomes hyperoxygenated.

Before long, however, the character of a pediculated interstitial tumour invaginating itself into an air-vesicle is destroyed, and the tubercle-nodule then comes to have a totally different character. The previously uniform free border which the tumour had (Fig. 57) now becomes destroyed, and the cells contained within the mass escape into the cavity of the air-vesicle. Similar alveolar thickenings form at different parts of the alveolar wall in the same way, and these opening into the air-vesicle produce great cellular distension of it. The capillary blood-vessels, which previously were clearly visible within the granulation mass, then become destroyed, so that the appearance presented by the tubercle in this stage is that of a group of three or more air-vesicles whose walls are only partially visible, and whose cavities are distended with large endothelial-like cells, with occasionally some blood-corpuscles.

It is in this stage that the tubercle growth somewhat resembles a catarrhal pneumonia, and it is now that it is so liable to be mistaken for an intra-alveolar or epithelial instead of an interstitial connective-tissue growth. The stage before this is not so easy to detect, seeing that by the time the disease proves fatal the most of the tubercles have advanced beyond it. It is this which evidently has in part misled M. Charcot, and has inclined him to believe in the identity of tubercle and catarrhal pneumonia. Both in primary and in secondary tubercle, however, and more especially in the latter, as we shall see, the distinction between the two diseases is so definite that no trained observer can possibly mistake them if sufficient attention be given to the subject. We shall have further proof of their non-identity when secondary tubercle of the lung is considered.

The tubercle growths having broken through the alveolar wall, the next thing capable of being observed is that the remains of the latter become incorporated with the large cells contained in the air-vesicles, so that a uniform rounded mass results, such as that represented in Fig. 51, in which the outlines of the original air-vesicles are barely recognisable.

In a catarrhal pneumonic nodule, on the other hand, the alveolar walls (Fig. 49) remain distinct, until destroyed by caseous degeneration. In this disease, also, the air-vesicles surrounding the nodule usually show more or less evidence of epithelial proliferation, but in tubercle the neighbouring air-sacs are comparatively free from any morbid change.

Besides growing in connection with the capillaries of the alveolar wall, it sometimes happens that the tubercle sprouts, parasite-like, from the inner coat of a large branch of the pulmonary artery. I have seen an instance of the lesion where the lumen of a moderately large-sized branch of the pulmonary artery was almost occluded in this way. The tumour seems to arise in connection with the arterial endothelium, but soon involves the tunica intima, leading to great thickening of it, and to the building up of a typical giant-cell tissue within it.

The origin of the giant-cells of tubercle has formed matter for great difference of opinion among pathologists since the time when they were discovered, as will be perceived from the following references :—

Virchow supposed that they were over-developed connective-tissue corpuscles (*Virchow's Archiv*, vol xiv. p. 51). Wagner stated that "he has several times seen a transformation of the branched cells of the (natural connective tissue) reticulum into the many branched and many nucleated giant-cells" (*Das tuberkelähnliche Lymphadenom*, Leipzig, 1871, s. 31). Klebs and Koster (*Virchow's Archiv*, vol. xliv. p. 286, and xlviii. p. 95), represented the giant-cells as transverse sections of capillary lymph vessels ; and Langhans (*Virchow's Archiv*, vol. xlii. p. 382) and Hering (*Histologische und experimentelle Studien über Tuberculose*, Berlin, 1873, p. 105), adopted this view. Schüppel (*Archiv, der Heilkunde*, 1868, heft 6) believed that they are formed by the confluence of several white blood-corpuscles within blood-vessels. Thaon (*Recherches sur l'Anatomie pathologique de la Tuberculose*, Paris, 1873) did not believe that there are such things as giant-cells in tubercle, but that they are merely bloodvessels filled with blood-coagulum! Curiously enough this most extraordinary view was supported by MM. Cornil et Ranvier (*Manuel D'Histologie pathologique*). Wagner and Rustitsky (*Virchow's Archiv*, vol. lvi. pp. 531, 532, and lix. pp. 217, 224) supposed that they are formed from cells of different kinds, but more especially from those of the vascular wall. Brodowski (*Virchow's Archiv*, vol. lxiii. p. 113) made out that they are formed in some way, not clearly explained, by "an abnormal functional capability of the vascular walls." Lastly, Klein (*Anatomy of the Lymphatic System*, II. "The Lung," p. 76) announced that they may be developed either from lymphoid cells ("displaced colourless blood-corpuscles") or from the alveolar epithelium. In the latter method of construction they may arise by the enlargement of a single epithelial cell or by the fusion of several.

There is little doubt that the giant-cells found in myeloid tumours, in ossifying bone, and in many other tissues, are all essentially of the same nature as the bodies which constitute so prominent a feature in the structure of tubercle. They have, practically speaking, the same appearance, and when carefully studied, it will be found that they pass through similar gradations in proceeding towards the evolution of the perfect tissue which they are destined to produce.

The first step in their development in primary tubercle of the lung is that one of the large connective-tissue elements which in great part constitute the young tubercle growth such as those depicted in Fig. 57, begins to enlarge more than those around it. Sometimes the enlargement is not limited to one cell, but two, three, or more of them simultaneously increase in dimensions. There is generally, however, one of them more developed than the others. In a short time the protoplasm of this cell becomes very granular, and it then forms what would be called a small giant-cell, or a large epithelioid or endotheloid body. It rapidly assumes colossal dimensions, so that in the course of a week to ten days it may have assumed all the characters of a fully grown giant-cell. The transformations that it subsequently goes through have already been described. In certain cases an appearance is sometimes seen, as Klein has described, as if several small cells ran together to form a larger cell. Such a method of giant-cell production is, however, more apparent than real, and if the matter be carefully studied the former method of formation will, I feel assured, be found to be the true one.

In order to get a clear insight into the significance of the giant-cells met with in tubercle, we must take a somewhat general view of the developmental relations of the connective tissues to each other. There cannot be the slightest doubt in the minds of those who have really worked at the subject, that the so-called connective tissues are all built up on the same plan, and that the differences in their outward configuration are merely superficial, and not of fundamental importance, being generally mere adaptations of structure to a special economic purpose. The different forms of white fibrous tissue, cartilage, bone, the cornea, &c., are all instances of tissues derived from the middle layer of the embryo in which the constructive model is the same, but in which certain modifications have occurred, subservient to the purpose for which the particular connective tissue is intended. In one, the matrix, or fully formed part of the tissue, is divided into fibrous bundles, when the particular purpose in view is to act as a means of attachment between two movable parts, as in tendon. In a second, as in bone, this matrix is rendered solid and rigid by impregnation with calcareous salts. In a third, of which cartilage is an example, it is so constituted as to afford an elastic counter-pressure where such is required; while yet, in a fourth, it is of such consistence as to allow light vibrations to pass easily through it; such being the structure of the cornea.

The materials employed in the building up of the connective tissues are chiefly two in number, first, a matrix, and second, nuclei, corpuscles, or cells, as they may be called, which lie upon it, and which subserve the purpose of keeping up its repair. Complicated as may seem the structure of some of the connective tissues, they can all be resolved into these two elements. The matrix is to be regarded as the perfect or formed material, the nuclei or corpuscles are the apparatus by means of which this is manufactured.

The basis substance or matrix is incapable of any further developmental progression, and the pathological changes which it is subject to are all of a retrograde nature. The nuclei, however, being protoplasmic in character, are continually liable to pathological changes under stimulation. To the effects of such stimulation upon connective tissues generally I must now direct the reader's attention.

The whole of the so-called "tumours" may be divided into two great classes. The first comprises those which originate in a structure derived either from the epi- or hypo-blast, and the second includes those which arise in some tissue developed from the meso-blast. The first is known as the class of epitheliomata, while the second is that of the simple histioid tumours and the sarcomata.

From the study of the sarcomata, and of some of the simple histioid tumours, we can derive most important knowledge regarding the significance of tubercle. A sarcoma is a cellular mass, arising from a connective tissue, in which the cells do not elaborate a matrix, but in which they remain in an embryonic condition. The types of cell met with are either round, spindle-shaped, or giant-cells. Such a division is certainly true when the tumour is fully formed, but in the commencement of the development of such tumours it is not so. I hold that in all of them the original type of connective-tissue cell is the giant-cell, and that the others merely represent a further stage in the evolution of connective tissues generally.

A sarcoma is nothing more than a connective tissue which has become over-stimulated from some cause. A special feature of the stimulation is its persistence, so that the irritant, instead of dying out after a certain period, as in an ordinary inflammation, is continuous and apparently accumulative. It looks almost as if the irritant was in reality derived from the sarcomatous elements themselves.

The effect of this stimulation upon the connective tissue is that the completion of a matrix from the cell can never be affected, apparently because the whole energy of the cells is expended in giving rise to a numerous progeny, instead of throwing out the formed material which would constitute the matrix. There is not sufficient time afforded for the secretion of this formed substance, the generative activity of the cell being too great. In this respect they resemble the primordial cells of the embryo, which within a few hours may have multiplied a thousandfold without actually elaborating any truly connective substance.

If careful examination be made of the connective tissue which is in process of being converted into a sarcomatous tumour, the first thing visible is the enlargement of its corpuscles. So great is this enlargement at first, that the corpuscles, in whatever situation the tumour may be situated, assume the characters of giant-cells. Many of them reach the size of the giant-cells of tubercle, but in others, division and consequent multiplication occur so soon within them that they are almost immediately converted into sarcomatous cells of smaller size. There is always, however, in all such tumours, *a tendency* for the connective-tissue corpuscles to revert to the giant-cell type, and hence I conclude that the giant-cell is the embryonic type of all connective-tissue cells. It is quite a mistake to suppose that sarcomata connected with bone are the only tumours which exhibit giant-cells. All sarcomata, in their early stages of development, show cells having the giant-cell characters.

In the so-called myeloid tumours of bone the giant-cells have more the character of accessories than that of the real cells of the tumour. The greater part of such tumours is made up of a spindle-cell tissue, and the giant-cells merely represent bone-corpuscles which have been set at liberty from their calcareous matrix. They very soon undergo proliferation, just as in all young sarcomata, and are converted into the spindle or small round cells of the tumour.

Hence it is that the myeloid cells are always found most abundantly in the parts of the tumour near the disintegrating bone. The spindle-cell mass, growing as it generally does from the periosteum, causes absorption of the bone matrix, the bone corpuscles are liberated, and then participate in the general stimulation, so that instead of keeping up the repair of the bone they revert to their embryonic type, and, enlarging, constitute the giant-cells of the tumour. The reason why they are more usual in tumours of bone than in others is, that the bone is only gradually destroyed, and the bone-corpuscles are consequently not all set free at the same time; whereas in an ordinary connective-tissue tumour the whole mass of connective-tissue in the part is suddenly irritated, and, by the time that it constitutes a tumour, the myeloid type of connective tissue corpuscle has given place to that of a higher stage of development, namely, the small round or spindle cell.[1]

In Fig. 58, for instance, a cell is represented, taken from a young primary sarcomatous tumour of the parotid gland, which was not connected with bone. Part of the tumour had undergone mucoid degeneration, and this cell was taken from a part in which the jelly-like mucoid was most abundant. The bulk of the cells in the tumour were of a round and spindle type, but those suspended within the mucoid were evidently under specially favourable circumstances for returning to an embryonic condition, and of developing this to its utmost. The jelly-like medium in which they existed no doubt allowed of their freely throwing out processes, and quite possibly supplied nourishment so abundantly as to stimulate them to unusual growth.

---

[1] I cannot agree with those who suppose that the osteo-klasts, or giant-cells, set free in such bone-tumours, are the cause of the absorption of the bone-matrix. I look upon them much more as the result of its destruction from totally different causes, and regard the giant-cells merely as the bone-corpuscles which have been liberated, and which have assumed their original embryonic characters. They subsequently disappear, in sarcomatous tumours, by proliferating, and by assisting in forming the round or spindle cells of which such tumours mainly consist.

190   TUBERCLE IN THE HUMAN LUNG.

The particular cell represented is a typical giant-cell, having branching processes, as in those found in tubercle. The periphery of the cell is the part from which the processes come off, and this presents a formed or fibrous appearance.

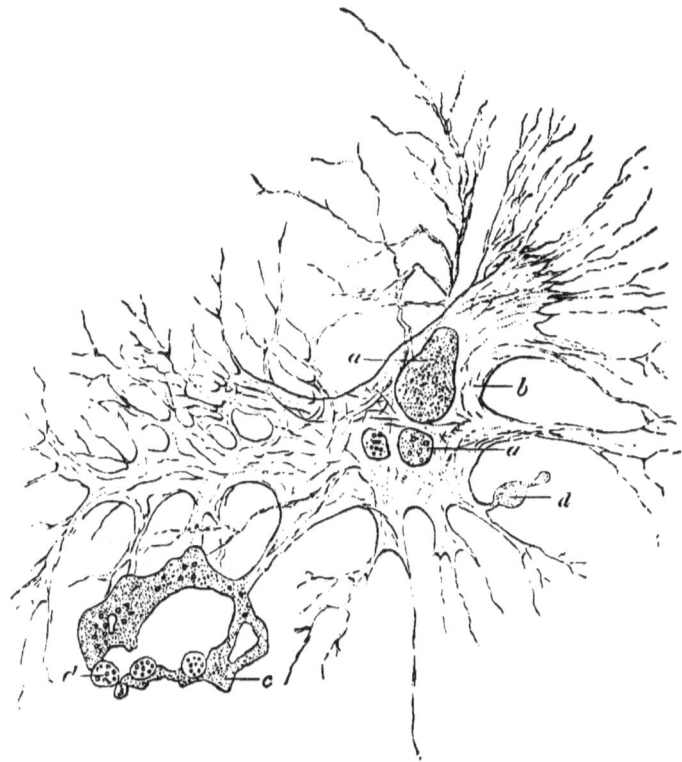

FIG. 58.—Enlarged connective-tissue corpuscle or giant-cell, taken from a young myxomatous sarcoma of the parotid gland, × 450 diams. *a, a*, nuclei of the cell ; *b*, the outer more formed portion, or periplast ; *c*, a portion of the periplast which has still retained its granular protoplasmic character ; *d*, ends or offshoots of the same.

In this enormous connective-tissue corpuscle, or giant-cell, I simply recognise a connective-tissue element which has been unduly stimulated, probably from an excessive amount of nutritive pabulum being supplied to it, and which, from lying in a suitable medium, has had every opportunity of developing to an unusual degree. Under such favourable circumstances it has become a giant-cell.

Fig. 59 shows another giant-cell, from a bone-tumour, and in this the close resemblance to certain of those seen in young tubercles will be apparent.

Now I will not go so far as to say that tubercle is a sarcoma, but I certainly have very good grounds for asserting that in an early stage of its growth it bears a very close resemblance to this class of tumours. There appears to be formed within the softening caseous source of infection some substance, possibly a ferment, which, being carried by the blood-stream into an organ, has the power of locally stimulating its connective tissue to great activity of growth. The result of this we have seen in the little mass of connective-tissue elements of which the tubercle, in an early stage of its development, consists, (Fig. 57). We have seen, further, that it is from the enlargement of one of these that the giant-cells are formed. That is to say, the portion of connective tissue in this stage is intensely stimulated, its elements become embryonic, and some of

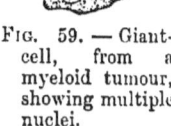

Fig. 59. — Giant-cell, from a myeloid tumour, showing multiple nuclei.

them reach the size of giant-cells, or what we have just described as typical embryonic connective-tissue corpuscles. In this stage, the tubercle is essentially a sarcoma.

The irritant, however, which has been carried to the part, and which has induced these changes, appears to be of an evanescent character. The stimulation which it excites dies out in course of time. On the supposition that it is a ferment, this would be expected; and then it is that the tubercle loses its sarcomatous character. The stimulation excited is excessive so long as it lasts, and hence the connective-tissue elements very rapidly revert to their embryonic type. It is, however, different from that which excites the proliferation in a sarcoma, in being merely of temporary efficacy, while the latter is permanently present, and appears to increase in intensity.

I have shown that the later changes in tubercle are essentially those of conversion of its cellular elements into fibrous tissue, and that its ultimate destiny is that of a little fibrous tumour. This quite coincides with the theory expressed, of the irritant being of the nature of a ferment, which excites great cellular proliferation while it lasts, but which, when worked out, loses this power, and allows the germinal cells to reach their ultimate development, namely, that of a fibrous tissue.

The reason why the tubercle-nodules are isolated at first undoubtedly is, that the irritant derived from the caseous destruction of a part is carried embolically into the organ, and merely excites the connective tissue locally. The fibrous ultimatum which the tubercles reach is much more common than might be supposed. It only requires the most trivial examination of a chronically tubercular organ to see how many of the tubercle-nodules no longer show any typically tubercular structure, but are converted into fibrous tissue. This will be more fully brought out when the subject of secondary tubercle is considered, and it will then be shown that many of the so-called cirrhoses of organs, both in children and in adults, are in reality the remains of a former tubercle eruption.

The conclusions which I have arrived at in regard to the significance of tubercle are :—

1st. That it is merely a form of connective-tissue growth.

2nd. That it is caused by an irritant acting upon the connective tissue, probably of the nature of a ferment, produced in the softening of a caseous mass.

3rd. That this is carried embolically into different parts of an organ, and stimulates them locally.

4th. That the tubercle at first has a close resemblance to a sarcoma, but that when the irritation has subsided, the connective-tissue elements organise and give rise to fibrous tissue.

5th. That the ultimate destiny of the tubercle-nodule is to produce a small fibrous tumour.

6th. That the presence of the giant-cells is merely an evidence of the return of the irritated connective-tissue elements to their embryonic type.

The difference between tubercle and a sarcoma is thus quite distinctly marked off, for while, at a certain stage of its growth, tubercle does somewhat resemble a sarcoma, yet its whole tendency, after the irritation has subsided, is to form fibrous tissue. In the sarcomata, of course, it is different, their great distinctive feature being that the cells do not reach full development, but remain in an embryonic condition.

The degenerations and complications of primary tubercle of the lung have not yet been considered; as, however, they will form a fitting sequel to the subject of secondary tubercle of the lung, their discussion will for the present be postponed.

On Secondary Tubercle of the Lung.

By the term "Secondary Tubercle" of an organ I mean, as previously explained, tubercle which has followed the softening of a caseous mass situated in the organ itself. The tubercle in such a case is the secondary disease, and is to be regarded in the light of a complication. In the lung there are two main causes of such local, infecting, caseous centres, namely, Interstitial Pneumonia or Cirrhosis with Bronchiectasy, and Catarrhal Pneumonia in the third stage. Both of these are accompanied by tubercular deposits; but as they are more abundant and less obscured by surrounding complications in interstitial than in catarrhal pneumonia, I shall specially consider secondary tubercle as occurring in the former.

On Secondary Tubercle accompanying Cirrhosis of the Lung.

When treating of cirrhosis of the lung as a sequela of bronchitis, it was shown that tubercle is common in this disease. The anatomical characters of the cirrhotic lung were at that time fully discussed, but the relationship of the tubercles to the surrounding interstitial increase was, for obvious reasons, merely hinted at. It will now be convenient, under the title of "Secondary Tubercle of the Lung," to describe more fully the special characters of the tubercles, with particular reference to the relationship in which they stand to the cirrhosis.

The clinical history is of great importance in understanding the pathology of this disease, as the symptoms often act as a guide in following the order of events. Instances must be familiar to every practitioner, although, curiously enough, it does not seem to be generally understood that it is tubercular. The reason is that it is a lesion of adult life, whereas the general supposition is that tubercle is essentially a disease of youth and childhood. This idea is totally fallacious, tubercle being a commoner disease in persons over twenty-one years of age than in those below it. Over a third of the cases diagnosed as ordinary catarrhal phthisis are, I find, examples of cirrhosis, and what I wish specially to emphasize at present, is the entire difference between it and pulmonary phthisis of catarrhal pneumonic origin. It seems to me that the physical signs accompanying it have not been clearly enough observed in the light of its being tubercular. The prognosis would require to be very guarded according as the one or the other disease is under consideration. During the last eighteen months I have made post-mortem examination of thirty-seven adults who died from what would ordinarily be called pulmonary phthisis, and, of these, fifteen were instances of tubercle accompanied by interstitial pneumonia, while the others were of catarrhal pneumonic origin. It is, therefore, clear that this affection is by no means rare, and its pathology, accordingly, deserves the closest attention.

As before mentioned, this is a disease of adults, and its onset is very insidious. A cough with slight bronchitic expectoration is usually first observed. The cough comes and goes, and is worst in winter. The patient loses flesh, and has a haggard look. There is slouching of the shoulders, hoarseness and marked retraction of the supra and subclavicular spaces. I am not aware that hæmoptysis is a characteristic symptom of this lesion. Slight attacks of pleurisy, accompanied or not by pleurodynia, and a general feverish condition, are noticeable features. The fingers become clubbed, and there is occasionally some œdema of the feet. These symptoms continue for several years, the patient not usually being incapacitated from following his employment. Some complication, often renal, at length arises, and this brings about the fatal result.

The post-mortem appearances are the following: the body is much emaciated, and the upper part of the chest has a retracted appearance, sometimes more on one side than the other. Flattening, both below and above the clavicles, is very evident. It is due to the shrinking of the lung substance, and to the traction of the cirrhosed lung upon the tissues of the lower part of the neck and the upper part of the anterior wall of the thorax.

The pleuræ are invariably united by fibrous adhesions, usually continuously, but at other times only at intervals. There may be fibrinous adhesion at one part, while the remainder of the pleural cavity is obliterated by fibrous union. At the same time the pleuræ are greatly thickened, and are coarsely fibrous and leather-like. The thickening is greatest towards the apex, and this no doubt contributes to the dulness frequently experienced on percussion in this neighbourhood. It may reach the extent of a quarter of an inch or more. If the visceral layer of the pleura is carefully examined several groups or rows of tubercle-nodules may be seen in it. They have the usual character of tubercle in this situation, and are grey and gelatinous.

The organ is shrunken, and its contour is irregular, from the retraction or collapse of some lobules and the overdistension of others. The disease is sometimes unilateral. When the lung is removed from the chest, shreds of the subpleural thoracic tissues may be seen adhering to the costal layer of the pleura. If percussed after removal, some parts of the organ give a hyper-resonant note, while others are dull. It feels hard and fibrous throughout, and nodules of small size may be felt lying in its interior.

When incised, several cavities of various sizes can be noticed. The largest of these are usually at the apex, and they may range in dimensions from a hazel-nut up to that of a small orange. They are invariably *bronchiectatic* in character, and, as their appearance and mode of formation have already been fully discussed in the article which treated of cirrhosis of the lung as a complication of bronchitis, it will be unnecessary to say anything further of them at present. They contain a quantity of pultaceous softened cheesy-looking material, composed of caseous catarrhal products.

The whole organ is beset with interlacing bands of dense fibrous tissue. Their density is always greatest around the bronchiectatic cavities, where they sometimes feel almost like cartilage. The lung-tissue is often so compressed between such a bronchiectatic cavity and the pleura that it is barely visible. The fibrous hyperplasia takes place chiefly in the situations where interstitial fibrous tissue is most abundant, that is to say, in the deep layer of the pleura, in the interlobular septa, and in the adventitious coats of the bronchi and branches of the pulmonary artery. The interlobular septa can be seen as thick cords running down from the pleura to the dilated bronchi.

Gumma-like nodules, varying in size from a mustard-seed to a pea, are sometimes seen in certain parts of the lung. They are round in shape or have a sinuous border, and are cream-yellow in colour and caseous in consistence. They are sharply circumscribed, and, like gummata, are generally situated in parts of the organ where the fibrous thickening is densest. When carefully examined, they are found to be portions of cicatricial tissue which have necrosed on account of obliteration of a small branch of the pulmonary artery, as in syphilitic gummata. The manner in which the small arteries become obliterated is illustrated in Figs. 32 and 33. The caseous matter of which they are composed sometimes softens in the centre, and from this, as an infecting source, secondary tubercles may be locally propagated.

An important point to consider at present is the disposition of the tubercles in such a lung. The situations they occupy depend upon the position of the lymphatic vessels radiating from the caseous sources of infection. In this respect they resemble, in their distribution, the course taken by tubercle of the peritoneal coat of the intestine adjacent to a caseous ulcer of the mucous membrane. The caseous matter is absorbed by the lymphatics and induces the growth of tubercle within them. Such being the case, it would naturally be expected that the tubercles should be most numerous near the sources of infection, and that they should diminish in number in a direction outwards from this.

Now, the sources of infection, in this disease of the lung, are undoubtedly to be sought chiefly in the bronchiectatic cavities above alluded to. Catarrhal products accumulate in these, they caseate and ferment, and the débris is absorbed by the surrounding lymph channels. The tubercles, accordingly, run in lines, or are collected in masses around the bronchiectatic cavities, owing to the distribution of the lymphatic vessels.

The bronchiectatic cavities usually originate at the apex, this being the region where the cirrhosis is greatest. The tubercles, which abound in this situation, are so incorporated with the interstitial tissue that their outline cannot be distinctly observed. In the lower parts of the lung, however, the cirrhotic new formation may not be so dense, and here the individual characters of the tubercles can best be noticed. Indeed, the tubercle deposits sometimes seem to be the only abnormality in the lower part of the organ, the cirrhosed tissue being confined to the upper lobe. They are of the size of a mustard-seed, not so large as a millet-seed. They are grey and gelatinous in appearance, and are inseparably adherent to the lung tissue. They may be cut out, and, if squeezed between two pieces of glass, feel like little masses of cartilage, and are with difficulty destroyed. Next to the immediate vicinity of the infecting caseous centres, the adventitious coats of the arteries and bronchi, the interlobular septa, and the deep layer of the pleura are their commonest seats. They follow, in fact (and this is significant in regard to the channels of conveyance of the poison), the course taken by inhaled pigment particles.

Although the anatomical features of the nodules are alike in primary and in secondary tubercle, yet their lines of distribution are quite different. For, whereas in the primary form they are scattered universally throughout the organ, in secondary tubercle they follow the course of the pulmonary lymphatic vessels contained in the periarterial and peribronchial sheaths, the interlobular septa, and the deep layer of the pleura. Very little of the infecting virus appears to be taken up by the blood-vessels. Tubercles are occasionally met with in the liver or kidney in such cases, but only in small numbers; whereas, in primary tubercle of the lung, where the blood-vessels are the means of dissemination of the caseous virus, many organs are simultaneously tubercularized. In the primary form the tubercles are not found more abundantly in the interlobular septa and pleura than in other parts, while the bronchial glands are by no means necessarily the seat of tubercular deposits. In secondary tubercle, on the other hand, the bronchial glands are invariably much enlarged. They contain many tubercles, usually placed, along with pigment particles, at the periphery of the gland, in the neighbourhood of the lymph sinus—again indicating that the lymphatics have been the means of transmitting the irritant. The first lymphatic glands appear, as with inhaled pigment, to have the power of arresting the further progress of the irritant, so that adjacent parts do not become tubercular. It is, as already stated, the exception to find other organs than the lung infected with tubercles in the secondary disease, clearly showing that its propagation must have occurred through channels having a limited area of distribution—that is to say, through the lymphatics.

The same localised distribution of secondary tubercle occurs in catarrhal phthisis, in which disease, as is well known, general tuberculosis is not often met with. When the caseous softening occurs gradually there seems to be much more liability to the débris being absorbed by the lymphatics than by the blood-vessels. There is usually a non-vascular area around a chronic caseous deposit which apparently prevents the softened mass being removed by the blood-channels, while the lympathic radicles are capable of taking up a small quantity of it. I have seen a lymphatic gland, in an instance of general primary tuberculosis which had suddenly caseated and softened, and in which the caseous matter had been rapidly absorbed. The wall of the cavity which resulted was covered by a plexus of congested blood-vessels, and, no doubt, these had been the means of removing the contents of the cavity, and of distributing them generally throughout the body. In a chronic softening there is not any such vascular plexus to be seen on the wall of the softening part, but a hard layer of caseous substance intervenes between the surrounding blood-vessels and the fluid contents at its centre. This apparently prevents the blood-vessels taking up the débris in any quantity.

Very different is it where a caseous gland, for instance, undergoes rapid and complete softening, and where the surrounding blood-vessels are in direct contact with the abscess-like cavity which results. Here the irritating virus is absorbed in large quantity by the blood-vessels, and is carried throughout the body in the blood-current, giving rise to a widespread eruption of tubercle. The more gradually the caseous accumulation and softening occur, the less likelihood apparently is there of general tuberculosis being excited.

The lung is congested in the cirrhotic form of secondary tubercle. The blood also has a bright red colour. The cause of the congestion is the difficulty experienced by the blood in passing through the cirrhosed and tubercular organ. The capillaries are pressed upon by the cicatricial tissue of the organ, and the circulation within them is consequently rendered more difficult. The right ventricle of the heart is always dilated or hypertrophied in such cases, for a similar reason. The bright red colour which the blood has is owing to hyperoxygenation, due to the delay in its transmission.

The bronchi contain catarrhal fluid, and expectoration is sometimes copious during life, in the later stages of the disease. Catarrhal expectoration is of course much more a symptom of this variety of tubercle of the lung than of the primary. The disease in this is usually a bronchial lesion to begin with, and the bronchiectatic cavities become a fertile source of catarrhal discharge as the disease progresses. The high tension of the circulating blood keeps the bronchial mucous membrane in a congested state, so that the epithelium can never be perfectly formed.

In Fig. 60 is represented a portion of a lung in an instance of the disease I have just described. It was taken from a part where the cirrhotic tissue was not developed to a very great extent, but where the tubercle nodules were abundant. Four tubercles are seen at *a, a, a, a ;* and, on comparing them with the primary tubercle represented in Fig. 57, it is evident that they are the same bodies, although their integral parts are more highly organised. It will be observed that the nodules are quite distinctly demarcated, yet that they are in part continuous with each other by intervening fibrous tissue (*b*). This is seen only where the tubercle growth has

TUBERCLE IN THE HUMAN LUNG.

been very chronic. It is never observed in the acute primary form.

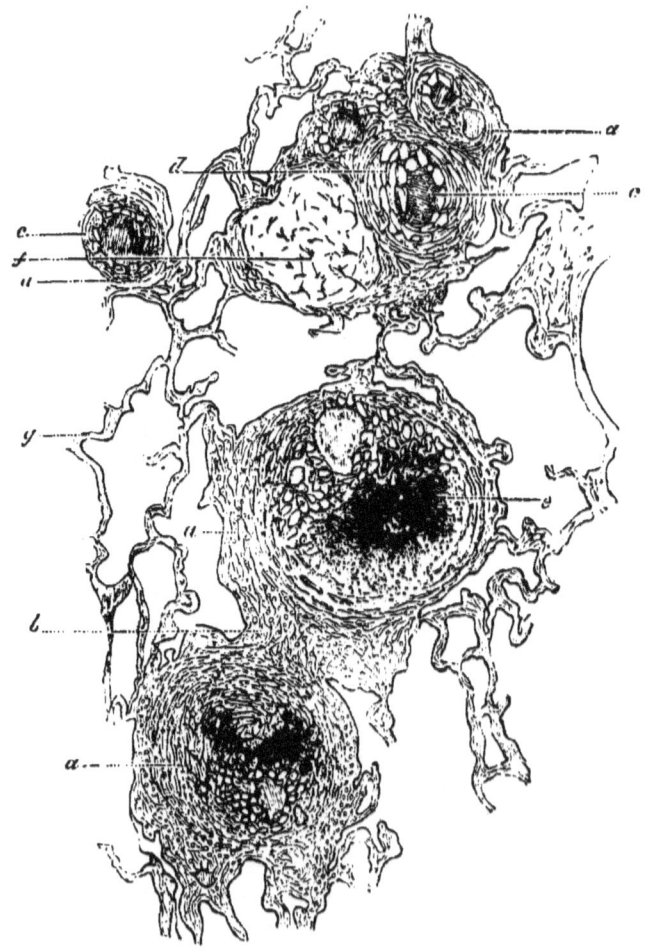

FIG. 60.—Secondary tubercle of the lung, ×50 diams. *a, a, a, a,* four tubercles ; *b,* thickened interstitial tissue uniting two tubercles ; *c,* giant-cells ; *d,* giant-cell reticulum ; *e,* centre of a tubercle caseating ; *f,* a giant-cell system which has become converted into a mass of fibrous tissue.

In all the nodules one or more large giant-cells can be noticed (*c*), from which there radiate processes, forming, by division and subdivision, the reticulum previously described.

The size to which the giant-cells may reach in secondary tubercle of the lung is truly enormous, so that, as will be seen from this drawing, they become objects easily recognisable with a magnifying power of fifty diameters. It can also be noticed that some of the tubercles, not all of them, are caseous in the centre (*e*). In one of the nodules the giant-cells and also part of the reticulum are caseous.

Masons who work on certain kinds of stone suffer from a peculiar disease of the lungs, due to the inhalation of stone-dust, the symptoms of which resemble somewhat closely those of ordinary phthisis pulmonalis. I have never examined the body of a stone-mason who has worked in the neighbourhood of Edinburgh without finding evidence of the effects produced by stone-dust inhalation. One of the features of the disease is that nodules, about the size of a millet-seed, or a little larger, are scattered over the pleura and throughout the lung-tissue, in the lines of the lymphatic vessels. They are extremely hard, sharply circumscribed, and occupy the same situation as secondary tubercles. When situated on the pleura, they are seen to be grey and cicatrix-like in the centre, and black at the periphery.

Now, when these deposits are examined microscopically, they are found to consist in great part of dense concentrically arranged bundles of cicatricial tissue, the particles of stone-dust lying in great numbers at the centre of the tumours and in the plasmatic spaces between the bundles. The stone-dust particles are very minute, and are either round or angular. They have a clear centre, and, when seen in mass, a greyish colour. They run in the course of the lymphatic vessels, and are occasionally accompanied by particles of carbon, which have been simultaneously inhaled. The cicatricial tissue is undoubtedly caused by the stone-dust irritating the fibrous stroma of the organ. Giant-cells are not found in these nodules, and the reason apparently is that the fibrous hyperplasia takes place gradually, so that abundant time is afforded for the nuclei of the stroma being organised. They do not rush into an embryonic existence, as in the case of tubercle, where the irritant is very acute, but pass through the stages of round and spindle cells very slowly, until the perfect fibres are produced. With this exception, however, the nodules in the stone-mason's lung are identical with tubercles. They in fact represent the ultimate development to which tubercles reach when sufficient time is afforded for their organisation. In both instances the nodules are fibrous-tissue formations, the one being the result of an irritant acting acutely upon the fibrous-tissue nuclei, and the other the product of a less virulent stimulant acting upon the same over a long period of time.

On the Mode of Development of Secondary Tubercle accompanying Cirrhosis of the Lung.

As secondary tubercle of the lung is of much slower growth than the primary variety, its mode of origin can be studied with exactitude. The first thing observable, in parts which are not as yet cirrhotic, is a little swelling on one side of an air-vesicle. This increases in dimensions, and then invaginates itself into the alveolus. It is by this process of invagination into the air-sac that space is afforded for the growth of the tubercle. Fig. 61 shows how such a tubercle originates from the wall of the alveolus. The walls of the air-vesicle are seen at d, and the tubercle nodule is noticed to occupy the greater part of the cavity. The nodule, as it increases in dimensions, becomes polypus-like, by drawing after it some of the alveolar structures, and these constitute a pedicle. The alveolar epithelium is also pushed forwards, and forms an investment for the tumour. The epithelial cells are represented in the figure lying over the tumour.

Not only does an isolated tubercle nodule grow in this way by invagination, but, when one has so arisen, many secondary out-growths may be produced, by the same invaginating process, from the original mass. Such conglomerations of giant-cell systems as those represented in

Fig. 60 are formed in this manner, and the extent to which the lung-tissue may be involved by such supernumerary growths is sometimes very great. A few tubercle nodules are first formed in a localised area, and then from their borders offshoots are projected, by a process of invagination, into the neighbouring air-vesicles. The lung becomes almost

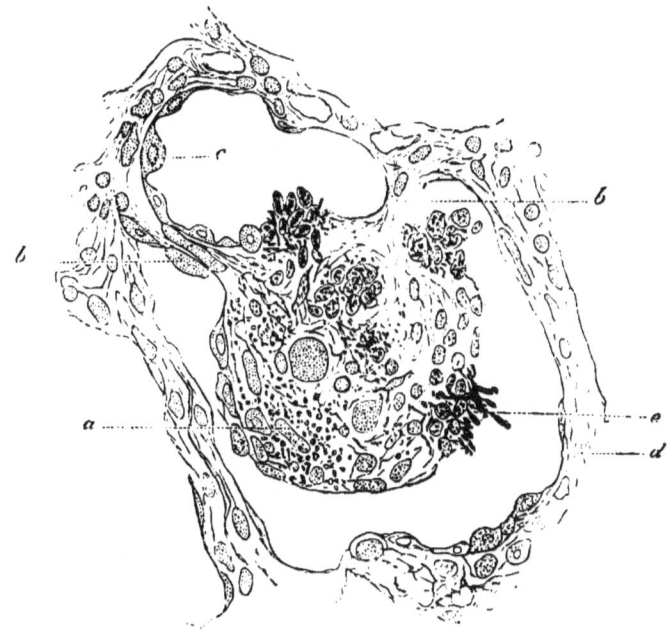

FIG. 61.—Secondary tubercle of the lung, showing how it invaginates itself into an alveolar cavity. *a*, the tubercle in an early stage of development; *b, b*, the alveolar wall; *c*, alveolar epithelium in a neighbouring cavity; *d*, the continuation of the alveolar wall; *e*, pigment particles, originally situated in the fibrous tissue of the alveolar wall, but which have now been carried forwards, and form part of the tubercle-nodule.

solid at such parts, from the immense masses of tubercles which are thus called into existence. In the course of time the tubercle outgrowth comes to fill the air-sac, and their walls become contiguous. I have not been able to make out what becomes of the two adjacent layers of alveolar epithelium. They disappear, apparently, by undergoing atrophy from the pressure of the tubercle growth; so that when the tubercle has reached its maximum size the alveolar walls and the substance of the tumour become united, what was formerly the alveolar wall now representing the boundary of the tubercle. Certain of the tubercles seen in Fig. 60 undoubtedly have been produced in this way, their sharp borders in fact representing what was formerly the wall of an air-vesicle.

In this respect, therefore, the growth of primary and of secondary tubercle, although in the former it is more rapid than in the latter, is alike. No doubt there are many tubercles in the secondary disease which do not follow this process of invagination, as, for instance, those which are placed in the centre of a very cirrhotic part of the lung, or in an interlobular septum. Here the growth is purely interstitial, and the tubercle involves merely the surrounding cicatricial fibrous tissue. But, where the tubercles lie apart from the cirrhotic portions of the lung, and here their development can be best studied, the manner in which they grow into the air-vesicle is practically the same as that seen in the primary disease. The true difference in their mode of development is, in reality, to be found in their points of departure, in the one case being the blood-vessels, in the other the lymphatics. Their subsequent course is alike.

THE DEGENERATIONS OF TUBERCLE.

The structure of a tubercle when fully developed has already been described, consisting as we have seen of a giant-cell fibrous network. In many tubercles, however, all the parts of the typical giant-cell system are not seen. The reason of this is that the tubercle is liable to degenerate, so that it is destroyed before or after being completed. The giant-cells are always present, but the reticulum, being a formation of later date, may not be noticed in certain instances.

TUBERCLE IN THE HUMAN LUNG.

The commonest degeneration to which tubercle is liable is the caseous. From the manner of its formation, necessitating as it does the destruction of the capillaries in the neighbourhood, the blood-supply of the tubercle is liable to be cut off at an early stage of its development. More especially is this true of primary tubercle of the lung, where the nodules arise from the blood-vessels. Such tubercles always show evidence of caseous decay, so that frequently the structure is arrested before it is complete. Secondary tubercle is of slower growth, and being of lymphatic origin is not so liable to implicate the vessels at an early period. Even here, however, after a time, many of the nodules become caseous (Fig. 60, *e*).

The granular appearance characteristic of caseation is always first perceived at the centre, and, when softening occurs, a microscopic cavity may result. Such cavities, however, do *not*, in the lung, run together to constitute larger ones, for, as soon as the centre of the tubercle has softened and has been absorbed, the peripheral fibrous part contracts, so as to obliterate the space which has been left. *I have never seen such a thing as phthisis pulmonalis resulting from caseation and softening of tubercle nodules*, and I do not believe that there is in reality such a thing as "tubercular phthisis," in the strict acceptation of the term. No doubt the gumma-like masses which have been previously described as complicating secondary tubercle of the lung, accompanied by cirrhosis, do sometimes soften and form cavities, but these are due to accidental obliteration of the blood-vessels, and are more an accompaniment of the cirrhosis, than the effect of the softening of the tubercles. It also sometimes happens that in instances of secondary tubercle of the lung accompanied by cirrhosis, there is a little localised catarrhal pneumonia in some part of the organ. The anterior margin and apex are perhaps the commonest situations of such. Softening and phthisical excavation of these catarrhal masses are frequently noticed, but this is not excavation from the softening and running together of *tubercle* nodules. Does a primary uncomplicated eruption of miliary tubercle such as one sees in children resulting from some distant caseous infecting source lead to excavation of the lung? I have never seen it. The tendency of the tubercles in such a lung, if the person lives long enough, is to become fibrous and to give rise to a cirrhosis of the organ. What I wish specially to emphasize is that uncomplicated tubercle of the lung does *not* cause excavation, and hence that the term tubercular phthisis, when used in this sense, is erroneous and misleading.

Secondary tubercle of the lung, on account of the cirrhosis which is associated with it, may lead, and often does lead, to bronchiectatic excavation. Such cavities are constantly mistaken on casual examination for cavities produced by dissolution of the lung-tissue, and the disease is called pulmonary phthisis. There is not in reality, however, in such a case, any destruction of lung-tissue from softening. The disease is essentially one of cicatricial contraction, leading to bronchial dilatation.

If, then, the tubercle caseates in the centre, what becomes of the caseous débris? The softening takes place locally, not *en masse*, as in a catarrhal pneumonia; and, as the caseous centre of the tubercle liquefies, it is entirely absorbed by neighbouring lymphatics, and is removed in this way. The periphery of the tubercle, which is the most fibrous part, meanwhile contracts; so that, as absorption of the centre takes place, the whole tumour, as before mentioned, shrinks, and thus prevents the formation of a cavity.

The caseous matter, in the course of time, is removed in this way by the lymphatic vessels, and is capable of again exciting tubercle growths within them at some distant part. Thus it follows that when tubercle is set up in an organ there is extreme difficulty in getting rid of it. For, once given a caseous centre, infinite numbers of generations of tubercles may originate from this, each individual tumour, as it caseates, propagating an offspring of young tubercles in its vicinity.

There is the closest analogy between the caseation of a gumma and that of a tubercle. A gumma is a caseous mass situated in the midst of a dense cicatricial tissue. It is a well-known fact that it may be absorbed, but that it has little tendency to form a cavity. The reason of this is that, as the softening and absorption of the caseous mass proceeds, the cicatricial tissue around shrinks, and prevents any vacuity taking place.

The other degeneration to which tubercle is liable is the *fibrous*. Indeed this can hardly be regarded as a degeneration, seeing, as previously described, that it is the ultimate destiny of all tubercles to become fibrous if they live long enough. It is of common occurrence, especially in *secondary* tubercle of the lung. It has previously been shown how the fibrous transformation is brought about—how the giant-cells and their reticular processes gradually become developed into fibrous tissue, and how, finally, not any trace of the previous giant-cell system is to be found. Fig. 60, *f*, shows a giant-cell system which has undergone this fibrous development. The other giant-cell systems, which are seen in the same tubercle at the side, are probably young offshoots from that which is fibrous, and in which, consequently, the fibrous degeneration has not as yet occurred. Such an appearance is very common in secondary tubercle of the lung.

Even such a tubercle (*e*) as that seen below the one which is fibrous (*f*), although at present caseous in the centre, would, when the caseous matter was absorbed, ultimately become converted into a fibrous mass, without any perceptible tubercle-structure. A circumscribed fibrous tumour, in which the bundles of fibrous tissue are fully developed, with plasmatic spaces lying between them, is what it becomes transformed into. Where, however, the centre has been destroyed by caseation, the rounded shape which the nodule originally had is lost, and an elongated fibrous thickening is all that remains.

The combination of secondary tubercle with cirrhosis of the lung has been sufficiently dwelt upon, and it was assumed that the cause of the cirrhosis is a chronic bronchitis, a catarrhal pneumonia, or some other local source of irritation.

There seems, however, very little doubt that *the tubercles* which form in the organ during the course of the disease are also a fertile source of fibrous overgrowth. Indeed, I look upon the presence of tubercles, in such a case, merely as a peculiar variety of fibrous hyperplasia. The whole character of parts of the lung which have not been primarily cirrhotic, but in which tubercles have developed locally, shows that this is the ultimate destiny of many of these tumours. The tubercles in such places are formed first, and a localised cirrhosis follows. An organ which is cirrhotic in one limited area to begin with may thus, by the fibrous transformation of the tubercles which are developed from it, become wholly beset with fibrous tissue, the tubercles in the course of time losing their characteristic shape, and giving rise to a widespread interstitial thickening.

The same thing is sometimes seen in the liver, more especially in children. It has often been shown that children suffer from cirrhosis of the liver. Several cases of the kind have lately been recorded in this country and abroad. Klebs draws attention to what he calls "miliary syphilitic new formations," which occur in the livers of children (*Handbuch der pathologischen Anatomie*, p. 440). He seems to assume that they are syphilitic because he cannot otherwise account for them. They occur, according to him, in the insterstitial tissue of Glisson's capsule. They are yellow in colour, and, he confesses, have a close resemblance to tubercles.

I have met with one or two instances of this disease, and have made very careful examination of them. I consider there is not the slightest doubt of the disease being tubercular. There is, however, this peculiarity about it, that it is accompanied by a diffuse cirrhosis, which, the longer the child lives, becomes more and more developed. A little careful study of the nodules easily convinces one that they are the source of the cirrhotic fibrous tissue. The tubercles are very thickly scattered through the organ ; they degenerate, or rather organise, into fibrous tissue ; one such fibrous mass joins with another ; and, in this way, an interlacing network

P

of fibrous tissue results, producing shrinking in the volume of the organ, as in an ordinary cirrhosis in the adult. In the course of time the tubercles may all disappear, and nothing but the cirrhotic bands arrests the attention. This disease of the liver in children is analogous to many cases of tubercle of the lung in the adult, for, in both, the tubercles lead to fibrous new formation. We are too much inclined to look upon tubercle as necessarily a fatal disease, without discriminating between its effects in an organ such as the liver and those in a part like the cerebral pia-mater. Tubercle of the pia-mater is a fatal disease merely because it affects a vital part, but where it occurs in an organ such as the liver, it causes no further acute disturbance than so many encysted parasites would in a similar situation.

What then becomes of those tubercles, where a fatal issue does not at once ensue? In children of a strumous constitution caseation and softening of glands is of very common occurrence without proving fatal. Some infecting material must be absorbed from these, and, being transported by the blood-stream into neighbouring parts, will give rise to tubercle formation in them. If, however, vital parts, such as the meninges, escape, tubercular deposits may take place in many other organs without necessarily causing a fatal result, usually with very little constitutional disturbance. The child lives, perhaps for several years, and I have very little hesitation in saying that the tubercles conduce by their fibrous organisation to a cirrhosis of the organ in which they are placed. Certain cirrhotic states of the organs of children are thus accounted for which otherwise have not any appreciable cause of origin.

Catarrhal Pneumonia: Third Stage.

Having now described and clearly defined what I mean by a tubercle, and having shown under what conditions tuberculosis arises, the reader will be better able to understand the description of the third or phthisical stage of catarrhal pneumonia than if it had immediately followed that of the first and second stages. The reason of this is that tubercle frequently complicates the third stage. So far as we have gone, we have seen that the air-vesicles first become distended with catarrhal cells, and then, that these, with the involved alveolar walls, caseate. We have now to consider what further changes occur in the caseous parts of the lung. What I call the third stage of catarrhal pneumonia is where the caseous masses formed in the second soften and give rise to phthisical excavation.

It is seldom that an adult dies in the second stage of the disease. There is, almost without exception, evidence of excavation at some, it may be a limited, portion of the lung, and this, it will be remembered, constitutes the third stage of the disease. Children, as before mentioned, frequently die in the second stage, probably on account of the disease running a much more rapid course in them than in adults. Even in the mature lung, however, where the disease may have advanced far into the third or excavating stage, there are always portions, usually the base, where the malady is still in the second or caseous stage. A whole lung simultaneously infiltrated with catarrhal masses of exactly the same date is rare. The naked-eye appearances of the organ in the third stage are the following:—

It has been pointed out that fibrous adhesion of the pleuræ is to be expected in the second stage ; sometimes a fibrinous adhesion at one part, and a fibrous at another. In the third stage of the disease, however, the adhesion of the pleuræ of the affected lung is almost invariably complete, and it is usually impossible to separate the one from the other, especially opposite a locality where a vomica exists. The thickening which may have taken place in the membrane is in some cases very great. If a portion of the pleura should happen to be non-adherent, numbers of grey, gelatinous, and rounded nodules can be seen in it, which are freely movable with the pleural membrane. When cut into, they are found to be confined to the pleura itself; they are not catarrhal pneumonic masses which are merely pushing their way through the membrane, but are veritable pleural structures. These bodies are tubercles of the pleura, and they are usually placed in its deep rather than in its superficial layer. They are caused by absorption of the tubercle poison from the cavities in the lung-substance, and their position in the deep layer corresponds to the course of the lymphatic vessels leading from the lung outwards. They are purely lymphatic tubercles, and in this respect resemble those seen in the peritoneal coat of the intestine in phthisis of that organ.

Situated on the vocal cords or adjacent mucous membrane of the larynx there are usually some similar tubercular nodules. In the tracheal mucous membrane such tubercular nodules are also occasionally seen, but not so frequently as on that of the larynx.

The mucous membrane of the bronchi is always very red and congested. Their lumina also contain a large quantity of yellow, tenacious, muco-purulent discharge, evidently partly of local origin, partly derived from the vomicæ scattered throughout the lung-substance. It is rare that the epithelium lining the mucous membrane is normal : it usually presents the appearances formerly described as characteristic of bronchial catarrh, either of an acute or chronic nature.

The apex of the lung, as is known, is the situation in which softening and destruction of the lung-tissue usually commences. The whole of the upper lobe may be converted into one large cavity, or there may be multiple smaller cavities scattered through it. The characters of these

THIRD STAGE.

cavities are quite distinct ; their shape is irregular, sometimes rounded, or, what is quite as common, cuneiform. Their walls are very rough and nodulated, the nodulated projections being portions of caseous lung-tissue which are in process of liquefaction. Bands of a fibrous nature and of varying thickness are seen running across the cavities, and, if their points of attachment be carefully observed, it will be seen that they pass from the deep layer of the pleura, on the one hand, down to the wall of a large bronchus on the other. These are generally described as being the vessels of the lung, dissected out by the dissolution of the surrounding lung-parenchyma. Now, although this is usually accepted as the explanation of these bands, it is obviously untrue ; for, in the first place, there are not any vessels of this size running from the wall of a bronchus with almost undiminished calibre to the deep layer of the pleura, and, in the second place, the pulmonary and bronchial vessels have a much more disjointed course, and, even when injected, could not be dissected out in this manner. The cords are very thick, and run so directly to definite points on the inner surface of the pleura that it is evident they cannot be branches of either of the above-named vessels. When they are microscopically examined their structure is seen to be that of dense masses of fibrous tissue, with a few blood-vessels of ordinary size contained within them, and when their course and distribution are studied there cannot be any difficulty in understanding what they are. They are simply the interlobular septa of the organ, now much thickened, which, by having more power of resisting the process of destruction going on around, have been left in an isolated condition when the infiltrated lung-tissue became disintegrated. They are abundantly supplied with vessels, and hence, probably, it is that they are less easily destroyed than the infiltrated lung-parenchyma, in which the blood-supply has been gradually although completely cut off.

The cavities contain more or less viscid yellow fluid, frequently with curdy masses contained in it—the remains of the necrotic lung-tissue. This fluid contains great quantities of granular matter and minute oil-globules. If, as not uncommonly happens, a large bronchus communicates with the cavity, then a certain number of cellular structures may also get mixed with it, being the elements of the catarrhal discharge thrown off from the bronchial mucous membrane. But, if the cavity forms a shut sac, the contents are usually merely granular debris and minute oil-globules, without any great abundance of cells. The fluid is, in fact, merely the result of the liquefaction of the caseous matter formed in the second stage, and it is not pus, as might from its appearance be supposed. Should a bronchiectatic be mistaken for a true phthisical cavity, then, no doubt, large numbers of epithelial and other cellular structures will be found in it.

The lung-tissue around these cavities is usually densely packed with caseous catarrhal masses, all more or less in a state of disintegration. Besides these, however, there are occasionally to be seen small, round, grey and gelatinous bodies, having an appearance in certain instances identical with, in others closely resembling, tubercles. Niemeyer, in his classical work[1] on the subject of pulmonary phthisis, has drawn especial attention to these, and concludes that they are tubercles in all cases. At the time his work was written the histology of tubercle was not so accurately known as it is at present, and hence, from the author not defining exactly what he means by the term, there is some difficulty in understanding how far this observation may be trusted.

There is not the slightest doubt that, in the neighbourhood of such phthisical cavities, giant-cell structures are sometimes seen. They lie in the interstitial tissue, and more particularly in the interlobular septa. They are, however, usually very small, and are so obscured by the surrounding dense infiltration, that it is almost impossible to see them with the unaided vision. The bodies that one sees so often in the surroundings of such cavities, and which have a very close naked-eye resemblance to tubercles, have not usually any

[1] *Klinische Vorträge über die Lungenschwindsucht*, Berlin, 1867.

THIRD STAGE.

true tubercular structure, but are merely isolated groups of air-vesicles filled with caseous catarrhal secretion.

Let us examine a little more closely the appearances presented by a catarrhal pneumonic mass when softening. It will be remembered that the catarrhal pneumonic patch is an accumulation, within certain groups of air-vesicles, of cellular structures derived from the alveolar epithelium. Both these and the walls of the air-sacs in which they are contained become granular and caseate. Previous to this the blood-supply to the affected group of air-vesicles had been gradually cut off, on account of the pressure exerted by the accumulating catarrhal cells. The parts which have caseated are dead, and therefore the key to understanding the cause of their softening and disintegration is possessed in the following problem : What will happen to a mass of dead, comparatively dry and compressed animal matter, if kept for a prolonged period within the tissues of a living animal, at a body-temperature, and in a somewhat humid atmosphere, where it is practically excluded, by reason of its density, from contact with external agencies?

It has been shown by M. Duclaux[1] that, in the maturation of cheese, which presents almost analogous conditions to those under consideration, the main decomposition which takes place is that certain of the albuminoids, insoluble in water, become soluble. If then, as seems extremely likely, the insoluble proteids in a caseous tissue become soluble in the course of time, the cause of the liquefaction of a caseous mass can be easily explained. The cause of the absorption of these products will be also thus accounted for. It can in addition be understood how oil-globules, which previously, on account of their being bound up with the insoluble albumins, were almost unperceived, become abundant, and how an entire dissolution and separation of these from the now solid albuminous matter follows. From this it seems very clear that the process of liquefaction of the caseous mass in catarrhal pneumonia is due merely to

[1] *Comptes Rendus*, lxxxv. p. 1171 ; 1877.

chemical causes, and is entirely unconnected with the mechanism of the part in which the necrosis has occurred. The question, therefore, as to whether a caseous deposit will go on to softening, or remain in its indurated state, is purely one of time, and of the chemical changes which ensue within it.

Before the softening takes place in the catarrhal mass its structure becomes very dense in the centre, and all traces of

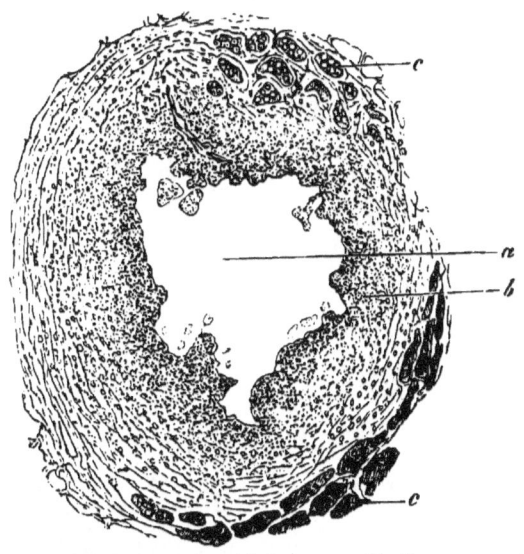

Fig. 62.—Catarrhal pneumonia, third stage, ×50 diams. *a*, the cavity formed by dissolution of the centre of the caseous pneumonic mass; *b*, the caseous edge; *c*, the compressed air-vesicles filled with caseating catarrhal products.

alveolar walls cease to be recognisable. Fig. 62 represents such a catarrhal pneumonic collection which has passed into the third stage. As will be noticed, there is a cavity (*a*) in the centre, whose ragged and granular edge sufficiently indicates its phthisical nature. The edge is undergoing gradual disintegration, as evinced by the remnants of the caseous matter which have become detached. The granular

THIRD STAGE.

degeneration, due to caseation (*b*), has advanced for a considerable distance outwards into the nodule, and has destroyed the contours of most of the air-vesicles. Further out, where the caseation has been less severe, the outlines of these (*c*) are still visible. The air-sacs are compressed and stretched round the cavity, and they are filled with the granular remains of catarrhal products.

When several small cavities are so produced, the tissue separating them also disintegrates, and then the one opens into the other and a larger vomica results.

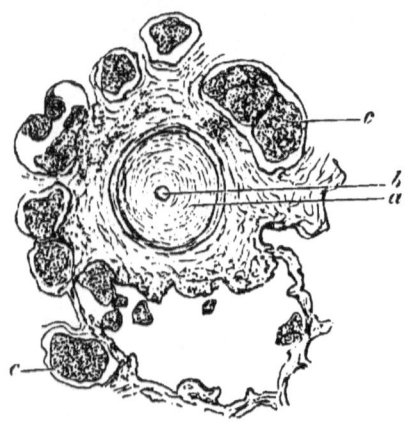

Fig. 63.—Catarrhal pneumonia, third stage, × 50 diams. Shows the obliteration which occurs in the branches of the pulmonary artery. *a*, thickened *tunica intima* of a branch of the pulmonary artery leading up to a cavity; *b*, the narrow lumen of the same; *c*, air-vesicles filled with caseous catarrhal secretion, and beginning to disintegrate.

It sometimes happens, however, that a considerable portion of a catarrhal lung caseates very rapidly, and breaks down almost like a slough. In such a case the obliterated condition of the blood-vessels leading to the part, such as that shown in Fig. 63, is apparently the cause. As has been demonstrated by Friedländer and others, an obliterative affection of the branches of the pulmonary artery, such as that seen in syphilis, is of common occurrence in phthisical lungs. The figure above alluded to illustrates this. It represents a portion of a lung taken from the edge of a phthisical cavity, and in its centre is shown a transverse section of a small branch of the pulmonary artery leading up towards the cavity. The lumen, or, rather, all that remains of it, is shown at *b*, its small size in all probability having completely prevented the passage of blood. The diminution in size is caused by the thickening of the *tunica intima* (*a*), whose inordinately great dimensions are apparent. Around the obliterated artery are several air-vesicles containing caseous catarrhal products, all in a more or less disintegrated condition. Now, when this obliterated state of the arterial branches supplied to an already infiltrated portion of lung becomes general over a wide area, it can easily be perceived how the diminished arterial supply will deleteriously act upon it, and tend to cause necrosis and disintegration *en masse*. This undoubtedly accounts for the presence of those large, rapidly formed cavities seen in certain phthisical lungs.

Small aneurismal dilatations of some of the vessels are occasionally found in phthisical cavities. They are usually about the size of a horse-bean, and by their rupture cause sudden death from profuse hæmoptysis.

The question of how far it is possible for a large phthisical cavity to cicatrise is, of course, one of the greatest importance. Were it not that the lung is in reality fixed to the costal wall by the adhesions which occur in the second and third stages of the disease, it might be possible for shrinking to occur to an extent sufficient to obliterate a cavity even of considerable size. Were the cavity situated in an organ such as the liver, for instance, it is conceivable, nay, likely, that closure of it would ensue by cicatrisation, if the debris were removed from its interior. The lung, however, is placed differently from the liver in regard to its surroundings. The costal wall being practically a fixed point, any cicatricial contraction in the lung-substance around such a cavity rather tends to widen than to diminish its interior. I hold it to be extremely questionable whether a large phthisical cavity ever becomes obliterated by this means. I have never seen any post-mortem evidence of it. Small cavities probably do heal by contraction.

If closure is not likely to take place in a large cavity, another beneficial circumstance may and does, notwithstanding, frequently occur. The caseous material being all expectorated or absorbed from the cavity, its wall becomes fibrous from surrounding cicatrisation, so as to render it no more dangerous a complication than a bronchiectatic cavity of equal dimensions. There is abundant evidence, both clinical and post-mortem, to show that patients may liv with such a cavity, or cavities, in the lung for many years. The great danger to such subjects is the further implication of the lung-tissue by the caseous pneumonic process, so that fresh areas of lung-parenchyma become infiltrated and destroyed.

The beneficial effects which phthisical patients experience from transference to an equable climate are, in all probability, due to there being little irritation caused by the inhaled air, and hence less liability to the excitement of catarrhal processes in sound portions of the organ. The lung has thus time afforded for cicatrisation of those parts already implicated.

ON A PECULIAR FORM OF CATARRHAL PNEUMONIA WHICH IS LIABLE TO BE MISTAKEN FOR TUBERCLE.

It has previously been stated that little reliance should be placed on mere naked-eye characters in determining whether a nodular deposit in the lung is tubercular or not. I propose, before concluding the subject of catarrhal pneumonia, to draw the reader's attention more particularly to this, as certain significant facts presently to be mentioned have an important bearing upon the elucidation of the etiology of many instances of diffuse tuberculosis.

A child has indefinite signs of catarrhal pneumonia,

passing, it may be, into those of general tuberculosis. The lung, after death, is found to be diffusely infiltrated with nodules, certainly sometimes a little larger than tubercles, but frequently as small or smaller. They have the same grey character at the periphery as tubercles, but are occasionally slightly yellow in the centre. Their shape is a little more irregular than tubercle, and in certain instances they tend to run together. Most of them, however, are quite isolated, and occur at intervals through the pulmonary tissue, very much as in primary tubercle of the organ. There is one point about this peculiar disease, however, which is significant. There is not any evident caseous source of infection in other parts, or in the lung itself. I have seen, certainly, tubercles in other organs in such cases, but these were evidently of later date, and corresponded to deposits secondary to those in the lung as an infecting centre. The history of such cases points to this being so, the meningeal or other tubercular disturbance being the climax of the disease.

The most curious point about these deposits is that they have not the slightest tubercular structure, but, in all respects, are identical with what is seen in the second stage of catarrhal pneumonia. They are small isolated groups of air-vesicles filled with epithelial products, the group invariably caseating in the centre. Every one has exactly the same appearance; there is not a vestige of any giant-cell structure; there is nothing of an interstitial character in the nodules. The whole process is one of catarrhal accumulation in the air-sacs, followed by necrosis of the mass; and the only difference between this and ordinary catarrhal pneumonia is in the fact of the nodules being small in size, isolated in character, and universally disseminated throughout the lung-substance.

I am well aware that, in the early stages of primary tubercle of the lung, there is a very close resemblance in the tubercle-nodule to a catarrhal pneumonic deposit. Still, in all these cases, where there is a distinct caseous infecting source, I have never failed to detect, in some of the older nodules, clear and undoubted evidence of the giant-cell structure. The tendency to organisation and isolation, which forms one of the most characteristic features of the tubercular as contrasted with the catarrhal pneumonic lesion, is also invariably present. The above, however, never show any tendency of this kind, and their periphery never has the

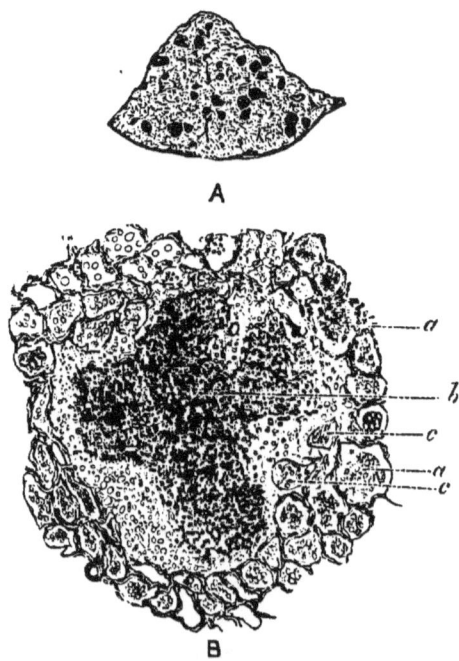

Fig. 64.—Disseminated catarrhal pneumonia. A, the naked-eye appearance of a portion of lung affected with this disease. B, one of the nodules from A, magnified 50 diams. *a, a*, air-vesicles filled with catarrhal cells; *b*, the caseous centre; *c, c*, the air-vesicles filled and becoming obliterated by caseation.

sharply circumscribed border, when seen microscopically, that the young tubercle-nodule has.

In order to show the characters of the two deposits, I have prepared two sets of drawings illustrating each. Fig. 64 (A) represents the naked-eye appearance of a small portion of a child's lung with the peculiar catarrhal pneumonic deposits in it, while Fig. 65 (C) gives the same view of a similar portion of a lung affected with primary tubercle, both of natural size. The close resemblance between them

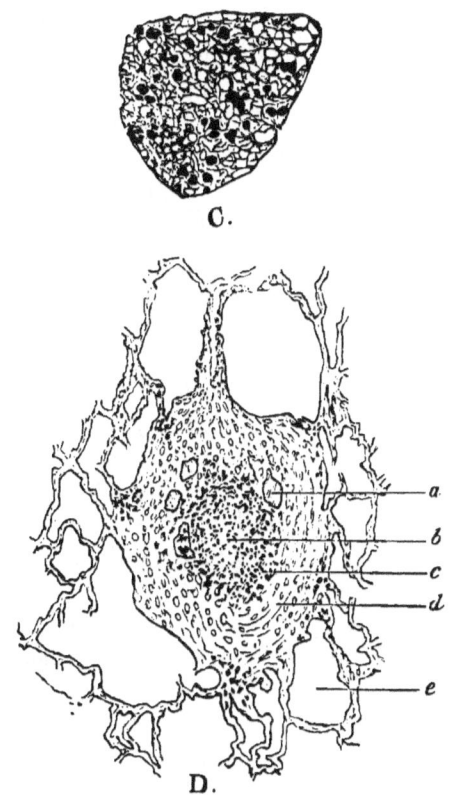

FIG. 65.—Primary tubercle of the lung. c shows the naked-eye appearance. D, one of the nodules from c, magnified 50 diams. *a*, giant-cell ; *b*, centre of tubercle becoming caseous ; *c*, pigment-particles contained in the nodule ; *d*, cicatricial periphery ; *e*, empty air-vesicles.

is apparent, so that, usually, with the naked eye it is impossible to tell whether we have to do with a truly tubercular deposit or not. This difficulty at once vanishes when they are examined microscopically, as will be seen from the subjoined drawings of each of them, magnified fifty diameters.

Fig. 64 (B) corresponds to the disseminated catarrhal pneumonia, while Fig. 65 (D) is the true primary tubercle. The difference between the two is evident. The catarrhal pneumonia is an infiltrated group of air-vesicles, while the other is an organised fibrous tumour with giant-cells abundant in it.

I had an opportunity of seeing how this disseminated catarrhal pneumonia commences in the lung of a child who suffered from acute catarrhal pneumonia after measles. There were the same miliary deposits, but in a stage previous to that in which they become caseous, and while the catarrhal secretion was as yet fluid. The wide dissemination of the catarrhal secretion seems to be due to inhalation of the catarrhal fluid during inspiration. The fluid secretion is widely scattered through the whole of the minute bronchial ramifications, and there is, apparently, not sufficient power to expectorate it. I have seen very much the same appearance produced by the sudden inhalation of blood from one of the primary bronchi. The disease, during life, presents some peculiar characters. It is usually diagnosed as disseminated tubercle of the lung; but, curiously enough, recovery from it is by no means uncommon. It must be a fact familiar to all physicians that a certain number of persons who suffer from what is diagnosed as acute disseminated tubercle of the lung make most wonderful recoveries, and, in a comparatively short time, are convalescent. It is in such cases that these disseminated miliary catarrhal deposits are found in the lung. It appears to be more fatal in children than in adults.

ON THE SUPPOSED CONTAGIOSITY OF TUBERCULOSIS
AND PULMONARY PHTHISIS.

Since Koch's famous paper (*Berliner klinische Wochenschrift*, No. 15, April 10th, 1882) on the contagiosity of tubercle was published the medical world has, so to speak, gone mad over the subject of tubercle in general and pulmonary phthisis in particular. The long expected solution of the phthisis problem has at last come! A bacterium has been discovered which explains everything!

Not only has this bacterium excited the greatest wonder and afforded the utmost consolation to the medical profession, but the public at large has been equally if not more impressed by it. Nay royalty itself[1] has deigned to look at this interesting organism, and seems to have been thereby immediately convinced of the infectiousness of phthisis : an imperial order is forthwith issued. Phthisical patients in military hospitals are to be separated from their brethren. Contact with phthisical persons is avoided. The consumptive in certain localities is shunned like a leper. The mere shaking hands with such is considered in some quarters to be "catching."

There is little doubt that in all this there is an instance of the *vivida vis animi* which so frequently accompanies the announcement of important discoveries, owing in great part to a misinterpretation of the facts. That Koch has made a discovery of importance no one will deny. It is the conclusions from it which to me seem so exaggerated and illogical.

By a special method of staining with certain anilin dyes he has found that there are bacteria in tubercular nodules and certain cheesy masses which give a distinct reaction. That is to say, other bacteria remain uncoloured by them, while the tubercle bacteria take on a vivid colour. He believes that it is tubercular nodules containing such

[1] See article by H. F. Formad, B.M., M.D., *Philadelphia Medical Times*, Nov. 18, 1882.

bacteria which alone are capable of being inoculated so as to produce a general tuberculosis.

They have a rod-like shape, are very slender, and from a quarter to half the length of the diameter of a coloured blood-corpuscle. They vary somewhat in dimensions, however, and may sometimes be actually as long as a coloured blood-corpuscle.

They closely resemble the lepra bacillus, with the exception that the latter is somewhat pointed at the extremities. The lepra bacillus, further, is stained with reagents which stain the nuclei of cells, while the tubercle organism is not.

They arrange themselves in groups or bundles often in the interior of cells, more particularly in the interior of the giant-cells of tubercle. At other times they remain free and isolated and many giant-cells are perfectly free from them. They are sometimes seen to be sporing, more especially when taken from an old phthisical cavity of the lung. They are most abundant in young tubercles, particularly those of a general acute eruption. In old tubercles they appear to be absent. They also abound in old phthisical cavities of the lung, but only in certain instances. Little masses, according to their discoverer, can be detached from these cavities, almost entirely made up of groups of such bacilli. Lastly, they are a common constituent of phthisical sputum, but, as in the phthisical excavations, they are not invariably present.

Bacteria of various kinds have been previously described as constituents of tubercle nodules. Thus Klebs (*Handbuch d. Path. Anatomie*) describes the granularity of the protoplasm of giant-cells as due to their being filled with micrococcus. Baumgarten (*Centralblatt*, April 15, 1882), claims priority in the discovery of the tubercle bacterium, but it is doubtful whether the organism described by him coincides in all particulars with that signalised by Koch. Aufrecht (*Pathologische Mittheilungen*, Magdeburg, 1881, and *Centralblatt*, April 29th, 1882,) in a similar manner described the centre of tubercle masses as composed not, as is generally

Q

supposed, of broken down cells, but of micrococci and bacilli. The same objection holds good here as in Baumgarten's observations, namely, that there is doubt as to whether such organisms as he describes are identical with what Koch named the tubercle-bacillus. Besides, I think that even the most ardent of mycologists will hardly be prepared to admit that the granular appearance of the centre of a caseous tubercle is owing to an accumulation of micrococci and bacilli.[1]

[1] [Koch's original method of staining the tubercle-bacillus was the following (*loc. cit.*). Since he described it he has given up this procedure in favour of that recommended by Ehrlich :—The object to be stained is either spread out on a cover glass dried and heated, or sections are made of it after hardening in alcohol. The cover glass, with the dried tissue or liquid upon it, or the section, is now laid for from twenty to twenty-four hours in the following solution.

Mix 1 ccm. methyl-blue with 200 ccm. distilled water. Shake them up repeatedly, and then, while still shaking, add ·2 ccm. of a 10 per cent. potash solution. This solution should not precipitate after standing for a day. If the staining solution is warmed up to 40° C. in a water bath the staining of the object can be accomplished in from half to one hour. The cover glass is then placed in a concentrated solution of vesuvin in water which must be filtered every time it is employed. When the tissue comes out of the meythl-blue solution it is much overstained. Washing it in the vesuvin removes the superfluous colour, and it then appears of a pale brown tint. By this method of staining, the tissue elements and broken down products appear of a brown colour while the bacillus stains of a beautiful blue. All other bacteria known to Koch, unless that of lepra, stain of a brown colour by this means. Sections of tubercular tissues when taken out of the methyl-blue solution are washed in distilled water, then laid in the vesuvin solution for from fifteen to twenty minutes ; afterwards washed in distilled water until the blue colour has all vanished and only a pale brown remains. They are then treated with alcohol, cleared in clove oil, and mounted in Canada balsam. A similar method of mounting may be employed when the tissue or sputum is dried on a cover glass.

Ehrlich's method (*Börners Deutsche Med. Wochenschrift*, 1882, No. 19), as employed by Dr. Heron (*British Medical Journal*, Oct. 14, 1882), is the following. Sputum is spread in thin layers upon cover glasses and allowed to dry in the air. It is then heated up to 212° F. for twenty minutes, or, what is as well, passed three or four times through the flame of a spirit-lamp ; a staining mixture is now made by adding 5 ccm. of aniline to 100 ccm. distilled water. The mixture is allowed to stand for twenty minutes, and then filtered through moist paper. To 30 ccm. of this mixture are added thirty drops of a saturated alcoholic solution of fuschine. A glistening metallic film should appear on the surface if the solution is of proper strength. If

Had Koch merely discovered a particular bacillus in tubercular and caseous tissues, his results would have lacked the necessary experimental verification for concluding that the bacillus in question was actually the agent which was instrumental in causing a tubercular eruption. It was necessary to isolate the organism, and then to utilise it for inoculation experiments. This he accomplished in the following manner. Serum from the ox or sheep is obtained pure, and is placed in a culture glass with a cotton plug in it. It is warmed daily for one hour up to a temperature

this does not occur, then more of the fuschine solution must be added until the film is clearly marked. The cover glasses, sputum downwards, are allowed to float for thirty minutes on the solution in a watch glass. On removal they have a deep red colour, and by far the greater part of the colour must now be removed by washing them in a mixture of one part nitric acid to two parts water. Then wash in distilled water. The tissue elements have lost most of their colour by this means, while the bacilli retain it. The former can be again stained by allowing a watery solution of methylene-blue to fall upon the cover. The superfluous blue colour is washed off with distilled water. The specimen should be examined, while still wet, with the microscope. If tubercle bacilli are present they will be seen as red rods upon a blue ground.

Another method which I have employed with great success is that devised by Dr. Heneage Gibbes (*British Medical Journal*, Oct. 14, 1882). The sputum is prepared on a cover glass as in Ehrlich's method. A staining solution is made of the following :—

Magenta crystals	grammes	2
Pure anilin	cc.	3
Alcohol, sp. gr. ·830	cc.	20
Aq. destil.	cc.	20

The cover glass with the sputum downwards is now placed on this as before, and allowed to remain for from fifteen to twenty minutes. It is afterwards transferred to a nitric acid mixture of the same strength as before, and allowed to remain until all the colour is removed, and then washed in distilled water. The staining of the tissue elements is accomplished by placing the cover in a saturated solution of chrysoidin for about five minutes. Finally, it is washed in distilled water, placed in absolute alcohol, dried in air, and mounted in Canada balsam. Dr. Gibbes states that sections of hardened tissues can be treated in the same manner with the necessary modifications, and the bacillus is shown by this method equally well in specimens hardened in spirit or chromic acid. In a later article (*British Medical Journal*, Oct. 21, 1882), Dr. Gibbes recommends a dilute solution of methylene-blue for staining the tissue elements in preference to chrysoidin. The staining reagents can all be had from Messrs. Becker & Co., 34, Maiden Lane, London.]

of 58° C. for six days. By this means the serum in most cases is sterilised and becomes converted into an amber-coloured jelly. This is then inoculated with the tubercular tissue taken from the lung of a tubercular animal just killed. In order to cultivate the bacillus the culture jelly must be kept at a temperature of from 37° to 38° C. During the first week nothing remarkable is noticed, but in the second week the growth of the organism commences. If some of the culture be inoculated, it produces an eruption of tubercle usually in about three weeks.

Such is a brief outline of Koch's remarkable observations, and against the conclusiveness of the experiments neither I, nor I daresay any one else, can found any serious objection. The only point of dubiety is that tuberculosis can be produced in rabbits and guinea-pigs from so many different causes, as, for example, the introduction of foreign bodies under the skin or into the abdominal cavity, the inoculation of the debris from a diphtheritic surface, the inhalation of different foreign matters, and so on, that really one becomes sceptical as to almost any procedure which consists in the insertion of an organic substance into their tissues. Many have, however, long held that an organism would ultimately prove to be the cause of tuberculosis. I myself in these articles, when they appeared in *The Practitioner*, avowed that a ferment produced in a caseous mass was the probable means by which a tubercular eruption was originated. Koch's discovery has therefore, like many others, proved the truth of what was formerly suspected. I shall therefore take it for granted that he has been able to isolate a bacterium which, when inoculated into animals of various genera, is capable of inducing a *tubercular* eruption. But this is all that I grant is proved by his experiments, and I hold that many of his conclusions, from the facts obtained, are entirely at fault and extremely misleading.

In the latter part of his paper he states that the discovery of the tubercle bacillus "renders it possible to limit the definition of what tuberculosis is, which formerly it was impossible to do with anything like certainty. There has always been a want of a definite criterion for tuberculosis, for one person includes therein miliary tuberculosis, phthisis, scrofulosis, bovine tubercle, and so forth; while another maintains, and with perhaps equal justice, that all these morbid processes are different. In the future it will not be difficult to define what is tubercular and what is not. Neither the special structure of tubercle, nor its deficiency in vessels, nor the presence of giant-cells will suffice, but the fact of its containing the tubercle bacillus, whether demonstrated by staining or by cultivation in sterile blood-serum, will be the one diagnostic point. Accepting this criterion, I must conclude in accordance with my researches that miliary tubercle, cheesy pneumonia, cheesy bronchitis, intestinal and glandular tuberculosis, bovine tuberculosis, and spontaneous and inoculated tuberculosis of animals are identical."

In a foregoing passage he makes this statement—that "the tubercle bacilli found in tubercular substances are not only the accompaniments of tubercular processes, but are the cause of the same." From what he says afterwards about the means by which the tubercle bacillus gains entrance to the body it is clear that he means to indicate that phthisis pulmonalis and the other affections above mentioned are, in common, due to infection with this bacterium. He accordingly calls all these processes tubercular, and would name a cheesy catarrhal pneumonic mass containing the bacillus in question a tubercle simply because the latter is present in it. A somewhat similar statement was formerly made by Cohnheim (*Die Tuberculose vom Standpunkte der Infectionslehre*, Leipsig, 1880), to the effect "that all is tubercular which by transference to properly constituted animals is capable of producing tubercle, and nothing is tubercular unless it has this capability" (p. 13). He ignores all structural appearances as a means of diagnosis, and includes such lesions as caseous catarrhal pneumonia, strumous glands, scrofulous joints, carious bone, and chronic strumous abscesses under the one generic term of "tubercle." They are all manifestations, he says, of the action of a poison, and are of the same nature as miliary tubercle-nodules. There is a poison within the system, of much the same nature as that of syphilis, which gives rise not merely to the eruption of tubercles, but also to the catarrhal pneumonia or other so-called strumous affection associated with it.

Let us see whether this is actually supported by clinical and pathological facts, and in doing so I shall draw upon illustrative examples which have come under my own observation, and whose pathology I have studied with care.

During an epidemic of measles, children, as is known to every medical practitioner, frequently suffer from acute catarrhal pneumonia. Nothing is commoner than for this catarrhal pneumonia to become chronic. It becomes chronic by caseating; and after death the lung is found to be infiltrated with caseous catarrhal pneumonia, and usually many of the different organs are beset with tubercles. In fact, death is frequently caused by the supervention of tubercular meningitis.

Now, if we look upon the catarrhal pneumonia as the result of the action of a tubercular poison, we are driven into the somewhat anomalous position of admitting that the tubercular poison and that of measles are alike. Interpreted in the light in which I view the subject of tubercle, and in that which was long ago promulgated by Virchow and Niemeyer, the catarrhal pneumonia is simply in keeping with the catarrhal affections of mucous membranes generally, so prevalent in measles. The caseation is a mishap due to insufficient blood-supply, from the undue pressure of the distended pulmonary alveoli upon the surrounding capillaries, and the tubercles in other organs are the result of maturation and absorption of the poisonous products. I cannot see that the pneumonia has any connection with the tubercle, otherwise than in the light of their being cause and effect. The tubercle in distant organs is due to a poison elaborated in the softening of the caseous substance; but the caseous catarrhal masses are not tubercle, although they may contain the tubercle poison.

A woman is delivered of a child. Peritonitis follows delivery. She partially recovers from this, but caseation ensues in the dried and accumulated peritonitic effusion. The caseous matter softens; it is absorbed, and general tuberculosis is excited. The symptoms of the deposition of the tubercles in the lung follow those of the peritonitis, and the final symptoms are those which betray the advent of tubercle in the cerebral meninges.

Is the peritonitic effusion, in such a case, the result of the application of a tubercle poison? Is not the logical conclusion rather that the peritonitis was primarily like any other peritonitis, only occurring in a puerperal woman, and that the bacillus which induced the tubercular eruption was grown in the softening caseous peritonitic products?

A person suffers from a decayed tooth; enlargement of some of the neighbouring lymphatic glands follows. They caseate, soften, and an eruption of tubercles follows. The same glands may previously have been enlarged from the same cause: it is when they caseate and soften that the general tuberculosis follows. Is it more reasonable to suppose that the enlargement of these glands was primarily due to a tubercular poison, or that it was the result of the irritation caused by the decayed tooth, and that the general tuberculosis was the response of the tissue to the stimulation caused by the absorbed poisonous products? If the caseation and the later-produced tubercles are one and the same thing, and if they are both the result of a constitutional poison, why is it that the one precedes the other, and why is it that the channels of dissemination of the tubercle-nodules radiate outwards from the caseous mass as a centre? It might be imagined that, if the caseous centre and the tubercles were manifestations of the same poison, they would appear simultaneously; but it is not so. The one invariably precedes the other. Nay, a mass of caseous glands, for instance, or any other caseous matter, may lie latent for months without producing tuberculosis, so long as they do not soften. It is when softening occurs that the danger arises. Show me that the tissue of such an enlarged gland is capable, on inoculation or otherwise, of giving rise to a general tubercular eruption before it has caseated, and then I shall believe that there is a specific poison at work in the production of its enlargement. But if it fail to do so until caseation has ensued, then I say that there is something evolved in this degenerative process which is the cause of the tuberculosis.

It is quite within the bounds of possibility, of course, that this peculiar irritant might find its way into the body, in certain rare cases, from without, as, for instance, from the contact of phthisical sputa. These are, however, quite exceptional cases, if they ever do occur. The bulk of the evidence indicates that the poison which excites the tubercle-eruption is generated in the tubercular subject, and is elaborated in the maturation of a caseous tissue.

There is, however, a very strong argument for the dissimilarity of the primary caseous deposit, and the tubercles developed from it, in the fact that they have totally different structures. As before mentioned, all tubercles, from whatever source arising, possess essentially the same histological elements and anatomical arrangement. But if we examine the infecting sources which give rise to them, they are found to have the most diverse composition. In the examples quoted, for instance, the infecting caseous source was, in one case, an epithelial accumulation in the lung, a fibrino-purulent effusion into the peritoneum, a lymphatic gland, and yet all were equally capable of exciting a tubercular eruption when absorption of the poisonous products took place. If, then, a catarrhal pneumonia resulting from measles, a fibrinous effusion, an enlarged gland, or an abscess, are to be classified under the generic name of tubercle, where, may I ask, are we to draw the line?

Cohnheim makes the statement that (*loc. cit.* p. 20), "*a tubercular or scrofulous product arises wherever the tubercular poison alights and remains for a sufficiently long time,* i.e. *until it has had opportunity of becoming imbedded in the part."* He states that by far the commonest means of the tubercular poison gaining admission to the human organism is with the respired air. Where from, however, he does not make quite clear. He believes that tubercle of the pleura and of the bronchial glands is due to the direct absorption of this inhaled poison by the pulmonary lymphatics, as in the pneumo-konioses. The trachea and larynx become secondarily affected, he states, from expectoration of the poison; but he does not say why the poison, in passing through these while being inhaled, does not produce in them more frequently a primary tuberculosis. The diphtheritic poison is arrested by the pharynx and larynx, and is deposited there first; why not the tubercular, if it is inhaled?

The same author is somewhat puzzled to understand how these cases of what he calls "isolated tubercular meningitis" are excited. He cannot explain, on the inhalation-theory, how the tubercular poison gets up to the brain-coverings and specially selects them. Mark, however, how this difficulty is got over. It is inhaled into the nasal passages, and then makes its way from them directly up to the base of the brain! Now, if this were true, why do the nasal membranes and *dura mater* escape infection, while the *pia mater*, which is disconnected in its blood-supply and otherwise from the skull and its membranes proper, is selected as the seat of the tubercular deposit? The truth is that he has confounded two things which, if taken separately, explain the whole pathology of such cases. There is, and I say it unreservedly, in those cases of so-called isolated tubercular meningitis, invariably, either in the brain-substance, the bone, or the encephalic membranes, or some other part, a cheesy deposit which is *not* primarily tubercular, but which usually has an ordinary inflammatory origin: it is this which acts as the infecting centre and which induces the tubercular meningitis. According to Cohnheim, however, this mere caseous mass is regarded as being throughout of the same nature as the tubercles which result from it.

Then, if the tubercular poison is conveyed by the respired air, why is it that attendants and officials in consumption hospitals do not habitually suffer from pulmonary phthisis? Statistics show that they are not more subject to the disease than other persons. Why is it that a phthisical person transferred into a household of otherwise healthy persons does not communicate the disease to them, if it is infectious in the ordinary acceptation of the term? The very freedom with which we permit of phthisical persons going about is proof that no distinct harm can be traced to the practice. The only circumstances under which I could conceive of such an accident occurring are where some of the sputum accidentally comes in contact with a part capable of absorbing it. In the case of husband and wife the opportunities for direct contagion, of course, are numerous; but even in them how seldom it happens that any ill effect is distinctly traceable to this cause. Is it not one of the commonest experiences that a husband or a wife may linger on for years with pulmonary consumption without inducing anything like infection, even although living in close intimacy during the whole of the time? Can the same be said of diphtheria?

It seems to me that in all this there is a want, on the part of those who hold these doctrines, of clear reasoning. Because we find the same bacillus in a caseous mass and in a miliary tubercle, are we, therefore, to conclude that they are alike, and that the bacillus is the cause of both? Such logic seems almost incomprehensible in men professing to have a scientific training. We might as well conclude that, because croton oil taken out of a glass bottle produced a pustular eruption when applied to the skin, therefore the croton oil was not only the cause of the eruption, but had also called into existence the glass bottle, and that both the glass bottle and the eruption were in their constitution alike.

So confused has the whole subject of tubercle and phthisis become, that one who has not specially worked at it no doubt will be in a state of perplexity what to make of all that has been written. After maturely considering the many points of argument, and as a result of my own experience, I have formulated the following conclusions which I think go as far as recognised facts will allow us.

1. Catarrhal pneumonia is a disease resulting from various causes acting as irritants upon the air-vesicles and small bronchi. There is not the slightest evidence to show that the tubercle bacillus floating about in the atmosphere is one of these.

2. The catarrhal products are liable to caseate in certain subjects, particularly those of a strumous constitution.

3. When they caseate there sometimes grows upon the caseous part a bacterium, which, if absorbed by the bloodvessels or lymphatics, is capable of irritating the tissues in which it becomes implanted, and of forming a little fibrous hyperplasia. This body and no other is a tubercle.

4. Other caseous masses throughout the body, more especially those connected directly or by means of lymph channels with a source of septic infection act as the means of cultivation. They are, particularly, strumous glands or abscesses which have opened externally, bronchiectatic cavities, caseous bronchial and mesenteric glands, and the strumous testicle where a sinus has existed.

5. These facts and many others seem to warrant the conclusion that the tubercle bacterium is originally allied if not similar to that found in a putrefactive wound, but has become specially modified by being cultivated on a caseous basis, so that when absorbed, instead of causing a slough of the tissue by its virulence, it merely irritates the tissue into forming a little cicatricial tumour. Tuberculosis and pyæmia, as Sanderson long ago pointed out, have therefore the closest relationship.

6. The tubercle poison, like the pyæmic, differs from that of syphilis and glanders in being generated *de novo* in a necrotic caseous tissue. That of syphilis and of glanders is propagated by contagion.

7. When once engendered, the tubercle poison can be transmitted from host to host by inoculation.

8. The reason of certain caseous tissues throughout the body not propagating a tubercular eruption is that by their position they are protected from extraneous contamination.

One of the questions taken up by Cohnheim in his paper is that of infection through milk derived from tubercular cows. The experiments of Orth [1] and others leave not the slightest doubt that rabbits fed on tubercular (Perlsucht) nodules develop tubercle in the pharynx, intestine, lung, and serous membranes, and certainly if the tubercle poison were capable of passing through the gland structures of the udder there is not any reason why the milk of a tubercular animal should not also excite tuberculosis. It is argued by Cohnheim because children are liable to caseous swelling of the mesenteric glands, that the latter are primarily tubercular, and that the cause of their tubercularisation is the absorption of tubercular milk from the intestine. He goes further, and states that "*the cheesy swelling of the lymphatic glands of the neck—what is generally known as scrofula—is due to the direct absorption of a tubercular virus with the food, and specially with infected milk.*" If this statement were really worth any attention, why is it that infants, whose diet is exclusively that of milk, and who are frequently suckled by phthisical mothers, do not develop tuberculosis? Tuberculosis is a rare disease in infantile life. How is this discrepancy to be accounted for?

In which are *tabes mesenterica* and scrofulous affections generally more common—in the inhabitants of our crowded cities, whose diet consists in small part of milk, or in those who live in country districts, and with whom milk is an article of diet at every meal?

Is there any evidence that the milk from tubercular cows has ever produced widespread or even isolated cases of tuberculosis in a community supplied with it? We know that scarlet fever and typhoid are sometimes distinctly traceable to an infected milk supply, but I have yet to learn that tuberculosis can be followed to a similar source.

[1] Virchow's *Archive*, lxxvi. 217.

The danger of contracting tuberculosis through the milk supply, although possible, seems, therefore, to be somewhat exaggerated. No doubt if the debris from a caseous cavity were to get into it, there might be danger of conveying the disease to the consumer; but that this regularly takes place in cows when they become tubercular seems improbable.

It has been frequently argued, however, that, as the tendency to pulmonary phthisis is hereditary, there must, therefore, be some poison inherent in the system in such cases which can account for its production. The whole question of heredity is, at the present day, in a most unsatisfactory condition, and there is no pathological subject in connection with which there is more misunderstanding. Before further considering it, I would wish to lay before the reader the following considerations :—

(a) Tuberculosis, and far less pulmonary phthisis, is practically unknown in new-born infants, and phthisical mothers do not give birth to tubercular or phthisical children.

(b) Statistics have shown that the more low-lying, damp, and wretched the dwellings of a community are, the more common is pulmonary phthisis.

(c) The members of a family who show a consumptive tendency are usually living under the same roof, and subject to the influence of the same surroundings. Individuals who are removed from these surroundings, and who are placed under better hygienic conditions, frequently escape falling victims to the disease.

(d) The more poorly nourished the subject is, the more liability is there to the production of phthisis.

(e) Persons who are attacked with catarrhal phthisis have weak circulatory organs, and consequently have a feeble power of propelling blood into a morbidly infiltrated part. The mere fact of a person having a powerful heart is almost a surety against the occurrence of caseous catarrhal pneumonia.

(f) The hereditary subjects of this disease are usually characterised by great epithelial development in the shape of hair. This occurs, sometimes, to an extraordinary extent, even on parts where hair is not naturally present in any great quantity, as along the spine.

(g) In children the pulmonary epithelium is of a much more germinal and, consequently, unstable character than in the adult; and catarrhal pneumonia is commoner in children than in middle life or old age. It is in those children who are weakly, and in whom the circulatory powers are deficient, that caseation of the catarrhal products is most likely to supervene.

(h) Those who suffer from valvular lesion of the heart, with cyanosis, do not subsequently become phthisical.

Now I hold that these facts form very strong premisses for an argument against the hereditary tendency of phthisis being due to a poison. If this poison-theory were true, why is it that the offspring of phthisical mothers or fathers are never phthisical, or tubercular, at the time of birth? In syphilis, where there is undoubtedly a virus at work, we see the effects of transmission in the offspring; but in phthisis all the evidence is absolutely opposed to that of a virus being directly communicated from parent to child.

The heredity of phthisis I would account for in a totally different way. Certain external circumstances, more especially a vitiated atmosphere and a damp locality, appear to be capable of engendering a peculiar habit of body, in which the pulmonary epithelium becomes much more sensitive to external impressions than it is naturally. This is what is termed a "delicate constitution," and it is characterised by the great liability such persons present to "catch cold" as it is called; that is to say, they are subject to slight attacks of catarrhal pneumonia. Such an unduly sensitive character of the pulmonary epithelium to outward impressions appears to be that which is hereditarily transmissible.

If, then, the vital powers of such a person be at a low ebb, should nutrition be ill carried on, and should, as usually happens, the circulatory powers be weak, we have to do with factors which will tend to cause the necrosis and caseation of any accumulated catarrhal products in the pulmonary chambers. The heart has little power in driving the blood through the infiltrated part; the blood itself is poor in nutritious constituents, and as a natural sequence the catarrhal products die and caseate. The same thing will happen in any tissue if we gradually cut off the blood-stream. A person with a powerful heart, on the contrary, is capable of driving blood through even such an accumulation, and so keeps up the vitality of the tissues until the obstruction is removed.

Over and above the hypersensitiveness of the pulmonary epithelium, however, there is evidence that in those who are hereditarily disposed to phthisis there is a tendency to exuberant growth of the epithelial tissues generally. This is shown by the profuse development of hair noticed in such persons on different parts of the body.

Altogether, after carefully considering the subject of the *constitutional* nature of catarrhal pneumonia, it seems to me very doubtful if we can pronounce it to be so in its commencement. Indeed it would be a matter for regret if such a conclusion were generally accepted. Every physician will testify to the fact of how much the treatment of the early stages of pulmonary consumption has been influenced for the better by the doctrines and logical inferences of Virchow and Niemeyer. The physician, resigned to the fatalistic view of catarrhal pneumonia being a disease due to a constitutional poison, regards the advent of a little hæmoptysis with pneumonic symptoms as indications that treatment of a radical nature is useless, and that all that can be done is in the way of amelioration. There could not be anything more prejudicial to the welfare of the patient, as the experience of every physician who adopts the opposite view will prove. The treatment of this disease entirely rests upon a correct appreciation of its pathology, and if we once recognise the principle that it does not become *constitutional* until the caseous masses have become septic by the growth upon them of a special bacterium, we have thereby a guide to treatment whose importance cannot be overestimated.

At the same time one cannot be too strongly imbued with the conviction that when the above-mentioned changes have occurred in the pulmonary accumulation the disease becomes one which is capable of exciting tuberculosis, either in the subject of it, or by contagion in an otherwise healthy person. Either through the saliva, the sputum, milk supply, or other vehicle, there is a possibility of its becoming inoculated and inducing tuberculosis. That such is of common occurrence in the human subject, I do not believe; but that there is the possibility of its happening cannot, I think, be disputed.

INDEX.

INDEX.

A

	PAGE
ADDISON, diapedesis	114
Air-vesicles, structure of	102
,, plasma-spaces	107
Arteries, chronic bronchitis	52
,, interstitial pneumonia	77
Arteriitis obliterans	76
Aufrecht, tubercle bacillus	225

B

	PAGE
BACILLUS of tubercle	224
,, ,, staining	226
,, lepra	225
Basement membrane	10, 27, 33, 35, 49
Bastian, cirrhosis of lung	69
"Battledore" cells	19
Baumgarten, tubercle bacillus	225
Bleeding	128
Blood, pressure of, in lung	110, 116, 129
Bronchi, structure	3
,, mucosa	3
,, epithelium	4
,, inner fibrous coat	11
,, muscularis	12
,, outer fibrous coat	12
,, in opium poisoning	18
,, congestion	21
Bronchitis, acute	24
,, ,, discharge in	25
,, ,, lung in	25
,, ,, first changes	27
,, ,, cellular infiltration	33
,, ,, lymphatic glands in	41
,, ,, recovery	43
,, chronic	44
,, ,, causes	44

INDEX.

		PAGE
Bronchitis, chronic, complications		69, 89
,, ,, from acute		44
,, ,, epithelium in		46
,, ,, basement membrane in		49
,, ,, inner fibrous coat		52
,, ,, muscularis		52
,, ,, arteries		52
,, ,, outer fibrous coat		55
,, ,, from valvular lesion		60
,, ,, from foreign matters		64
,, ,, from Bright's disease		68
Bronchiectasy		80, 95, 197
,, causes of		99
Brown induration		131
Buhl, on desquamative pneumonia		63

C

		PAGE
CARTILAGES, bronchial		56, 81
,, Filz on		58
Caseation, of tubercle		206
,, of catarrhal pneumonia		153, 154
Catarrhal pneumonia		99
,, ,, first stage		136
,, ,, second stage		147
,, ,, nodules in		148, 153
,, ,, caseation in		153, 154
,, ,, râles in		157
,, ,, pleurisy in		159
,, ,, hæmoptysis in		161
,, ,, third stage		211
,, ,, ,, pleura in		211
,, ,, ,, bronchi in		212
,, ,, ,, cavities in		212
,, ,, ,, aneurisms in		218
,, ,, peculiar form		219
Catarrhal cells		139
Cavities, phthisical		212, 218
Charcot, tubercle		171
Chronic interstitial pneumonia		69
,, ,, ,, tubercle in		70, 79
,, ,, ,, pleura in		70, 72
,, ,, ,, lobular septa		74
,, ,, ,, air-vesicles		74, 82
,, ,, ,, bronchiectasy		80
,, ,, ,, lymphoid deposits		82
,, ,, ,, causes of		82
Cheese, making of		155
Cicatrisation, phthisical cavities		218
Ciliated epithelium, formation of		19
Climate, benefit of		219
Coal-miner's lung		64
,, ,, ,, inhaled particles		64
,, ,, ,, bronchi in		64
,, ,, ,, pulmonary artery in		66
,, ,, ,, pleura in		66
,, ,, ,, course of inhaled particles		66

INDEX.

	PAGE
Cohnheim, diapedesis	114
,, infection from tubercle	229
Collapse	89, 92
Columnar epithelium, shedding of	28
Complications of bronchitis	69, 89
Congestion, acute, of bronchi	21
Connective tissue corpuscles	37
Connective tissues	186
Contagiosity of tuberculosis	224
Cornil, epithelium of air-vesicles	75
Corrigan, bronchiectasy	85
Croupous pneumonia	111

D

DEBOVE'S membrane	6
Degenerations of tubercle	205
Diapedesis, explanation	121
Discharge, acute bronchitis	25
Disseminated catarrhal pneumonia	219

E

EMPHYSEMA	89
Endothelium, changes in bronchitis	37
Epithelium of bronchi	4
,, ,, regeneration	17
,, ,, Schulze on	18
,, ,, origin	19
Expectoration in acute bronchitis	29

F

FIBROUS degeneration of tubercle	208
Filz, on bronchial cartilages	58
Fox, on chronic pneumonia	69
Franzosen-Krankheit	179
Friedländer, alveolar epithelium	75
,, arteries	76

G

GAIRDNER, emphysema	89
Ganglia, acute bronchitis	41
Germination of bronchial epithelium	32
Glands, mucous	15
,, ,, in bronchitis	38, 58
,, lymphatic, in bronchitis	41
Giant cells	172
,, ,, destiny of	176
,, ,, views as to nature of	184
,, ,, origin of	185
Gummata in cirrhosis of lung	196

INDEX.

H

	PAGE
HÆMOPTYSIS	161
Hæmorrhage, acute bronchitis	41
Heidenhain, mucus-forming cells	39
Heubner, arteries	76

I

INNER fibrous coat of bronchi	11
,, ,, ,, chronic bronchitis	51
Interlobular septa	12
Introduction	1

J

JAMES, on dropsy	120
Jenner, emphysema	90
Jones, inflammation	115

K

KLEBS, tubercle	209, 225
Klein, pseudostomatous cells	18, 105
,, lymphatics of lung	67
Koch, tuberculosis	224

L

LAENNEC, bronchiectasy	85
,, tubercle	149, 158
Lepra bacillus	225
Letzerich, bronchial epithelium	6
Liquor amnii, effect of, on epithelium	23
Lithosis	202
Liver, tubercle of	209
Lymphadenoid deposits	14
Lymphatic glands, acute bronchitis	41
Lymphatics of lung	14

M

MEASLES, bronchitis in	26
Mendelssohn, emphysema	90
Mucosa of bronchi	3
,, ,, thickening of	46
Mucous corpuscles, origin of	20
,, glands	15
,, ,, germination of epithelium	33
,, ,, in bronchitis	38, 58
Muscularis of bronchi	12
,, ,, in emphysema	54
,, ,, in bronchiectasy	54
,, ,, infiltration of	36
Myeloid tumours	189

O

	PAGE
Opium poisoning, bronchi in...	18
Orth, tubercle ...	236
Outer fibrous coat of bronchi...	12
„ „ „ „ infiltrations of ...	37
„ „ „ „ in chronic bronchitis ...	55

P

Part II. ...	100
„ Introductory ...	100
Perlsucht ...	179
Pleurisy ...	211
Pseudostomatous cells ...	18

R

Ranvier, epithelium of air-vesicles ...	75
Recovery, acute bronchitis ...	43
Regeneration of epithelium ...	17
Remak, bronchial ganglia ...	41

S

Sarcomata ...	187
Schulze, on epithelium of bronchi...	18
Socoloff, on bronchitis ...	30
Stewart, on bronchiectasy ...	97
Stirling, on bronchial ganglia ...	41
Stone-mason's lung ...	202
Structure of bronchi ...	3

T

Tabes mesenterica ...	236
Tracheo-bronchitis ...	24
Tubercle of lung ...	162
„ „ infection in ...	163, 197
„ „ in oxen ...	179
„ „ in rabbits and Guinea-pigs ...	180
„ „ conclusions as to nature ...	192
„ „ degenerations ...	205
„ primary ...	164
„ „ structure ...	172
„ „ giant-cells ...	172
„ „ origin of ...	180
„ „ resemblance to catarrhal pneumonia ...	183
„ secondary...	193
„ „ accompanying cirrhosis ...	194
„ „ bronchi in ...	200
„ „ congestion in...	200
„ „ mode of development ...	203

INDEX.

V

	PAGE
VESSELS, distension of, in acute bronchitis	27
Villemin, on tubercle	163

W

WHOOPING cough, bronchitis in	26
Williams, on caseation	179

Z

ZIMMERMANN, diapedesis	114

THE END.

LONDON: R. CLAY, SONS, AND TAYLOR, PRINTERS.

www.ingramcontent.com/pod-product-compliance
Lightning Source LLC
Chambersburg PA
CBHW021351230426
43666CB00006B/490